Performance Appraisal of School Management

Performance Appraisal of School Management

HOW TO ORDER THIS BOOK

BY PHONE: 800-233-9936 or 717-291-5609, 8AM–5PM Eastern Time

BY FAX: 717-295-4538

BY MAIL: Order Department
Technomic Publishing Company, Inc.
851 New Holland Avenue, Box 3535
Lancaster, PA 17604, U.S.A.

BY CREDIT CARD: American Express, VISA, MasterCard

PERFORMANCE APPRAISAL OF SCHOOL MANAGEMENT

EVALUATING THE ADMINISTRATIVE TEAM

DONALD E. LANGLOIS
Associate Professor of Educational Leadership
Lehigh University, Bethlehem, Pennsylvania

RICHARD P. McADAMS
Superintendent
Octorara Area School District, Atglen, Pennsylvania

TECHNOMIC
PUBLISHING CO., INC.
LANCASTER · BASEL

Performance Appraisal of School Management

a **TECHNOMIC**® publication

Published in the Western Hemisphere by
Technomic Publishing Company, Inc.
851 New Holland Avenue
Box 3535
Lancaster, Pennsylvania 17604 U.S.A.

Distributed in the Rest of the World by
Technomic Publishing AG

Printed in the United States of America
10 9 8 7 6 5 4 3 2 1

Main entry under title:
   Performance Appraisal of School Management: Evaluating the Administrative Team

A Technomic Publishing Company book
Bibliography: p.
Includes index p. 165

Library of Congress Card No. 91-67571
ISBN No. 87762-892-0

To Carol and Pat

Chapter Eight: **RELATING PERFORMANCE
APPRAISAL TO ADMINISTRATIVE
COMPENSATION**   **131**

Chapter Nine: **FROM THEORY INTO PRACTICE**   **155**

IN an educational era when terms like the *open classroom, modular scheduling,* and *affective education* have quickly faded from favor, the concern with "accountability" remains. The concept, however, has evolved. The initial concern was with assessing student outcomes. Next, we woke up to the need to improve the evaluation of teachers. This book completes the circle, effectively addressing the appraisal of administrators.

The completion is more than a raising of consciousness. Langlois and McAdams not only explain the why, but they also demonstrate the what and the how. After setting the stage by acknowledging the context that constrains and challenges us, they sequentially establish the priorities and elements of effective administrative appraisal. Then, in the heart of their handy volume, they describe and illustrate the procedures, policies, and forms for conducting appraisal of building-level administrators, the central office team, the superintendent, and, oft forgotten in this regard, the school board. Finally, rather than leaving administrative appraisal in an abstract vacuum, they appropriately tie it back to compensation arrangements and set up a positive process for realistic implementation.

Delightfully, too, the language is clear, concise, and crisp rather than "Pedaguese" that plagues the academic arena of our profession. Moreover, their attitude is consistently positive; they acknowledge and warn of the pitfalls, but focus on progress rather than problems. In the second chapter, for example, they paint this picture of the superintendent's affirmative duties as periodic meetings with building-level colleagues: "The superintendent listens, teaches, prods, suggests, paves the way, corrects, and applauds successes." In the third chapter they warn against being trapped in "administrivia" and "administrative gridlock." And in the fourth, they recommend *and reflect* "a relaxed yet business-like atmosphere."

The special strength of this book is its direct basis in the combined experience of the authors. Between the two of them, they have over a decade of experience spanning elementary, middle, and high school administration and over three decades in the superintendency. Drawing on this wealth of experience, they weave in the direct, relevant knowledge of not only the symptoms of, but also the solutions to, limiting accountability systems to students. They do not dwell on "war stories," but do offer enough examples to keep the book healthfully rooted in reality. The tone is practical as well as positive, but theory and research are not ignored: instead, theory is blended with practice in straightforward explanations and illustrations, and research is integrated into recommended policies and procedures, such as those for school board appraisal. As a supplement rather than substitute for their summaries and samples, they provide recommended readings for further details.

The authors appropriately acknowledge that "performance appraisal, done correctly, is hard work." But they show us that extending formal and functional accountability to all levels of leadership is both feasible and essential for true educational excellence. Practicing what they preach, they are, in effect, our coaches, giving us direction yet discretion, to incorporate job descriptions, individual administrative objectives, subordinate input, and self-evaluation into a streamlined, meaningful system for improving not only students, but ourselves.

The next time that I read a court decision such as *Jones v. Superior Court*, 233 Cal. Rptr. 464 (Ct. App. 1986), where an administrator's termination is judicially reversed because she received no evaluation of her performance, I shall be reminded of the clear and cogent lessons of Langlois and McAdams. Happy reading — and doing.

—Perry A. Zirkel
University Professor of Education and Law
Lehigh University

THE gap between theory and practice in school administration widens to a chasm in the area of administrative performance appraisal. This book, *Performance Appraisal of School Management: Evaluating the Administrative Team*, provides a workable blueprint for converting existing performance appraisal theory into successful practice. The approach to the subject includes a brief survey of the recent history of performance appraisal of school administrative personnel. Also featured are an analysis of the context for administrative performance appraisal in the 1990s and the presentation of specific, practical, and powerful models of administrative performance programs and practices.

A recurring theme in the book is the critical and pivotal role played by the district superintendent in performance appraisal. The book presents a more comprehensive approach to evaluation of the administrative team than is generally found in the literature on administrative appraisal. A major thrust of the recommendations which follow is the need for performance appraisal for all levels of the administrative team. The administrative team is defined as consisting of site-level administrators, central office administrators, the superintendent, and the school board.

The superintendent, by virtue of his or her position in the school system hierarchy, is the only person who can ensure that a high quality appraisal program is functional within the school district. Such a program includes systematic appraisal of all administrative subordinates to the superintendent, the superintendent, and the school board. The current scarcity of broad-based appraisal programs in most school districts is a serious deficiency which should be a matter of concern and remedial action within these school districts.

Only the superintendent is in a position to develop, implement, and maintain a rigorous administrative team appraisal process. School board members most often have neither the expertise nor the inclination to evaluate critically their own performance or that of their administration. Site administrators typically exhibit the normal human propensity to avoid strict accountability for their performance. Superintendents themselves often prefer to avoid the discomfort associated with a critical examination of their subordinates or themselves. Graduate school programs in educational administration have largely neglected this important dimension of school governance. Thus the typical school superintendent has neither the training, experience, nor inclination to pursue aggressively the development of a sound administrative appraisal program for the school district.

The first chapter of the book outlines the context of school administration in the 1990s. The social, cultural, and economic forces influencing the recruitment and retention of school administrators is reviewed in some detail. The potential impact of current educational initiatives such as school-based management, national testing, and schools of choice is also discussed. Chapter Two introduces several performance appraisal priorities of the authors that will be applied in later chapters to each level of the administrative hierarchy. Newer appraisal concepts such as coaching and staff involvement in administrative evaluation are discussed in detail.

Chapter Three introduces and explains the major elements of an effective performance appraisal program for the administrative team. This chapter thoroughly outlines critical appraisal elements such as the Job Description, Individual Administrative Objectives, subordinate evaluation of the administrator, and self-evaluation by the administrator. Chapters Four through Seven specifically apply the major elements of the proposed performance appraisal model to each level of the administrative team. Chapter Four applies the system to building-level administrators. Chapter Five considers those central office administrators who are generally considered to be a part of the superintendency team. In Chapter Six the model is applied to the evaluation of the superintendent. Chapter Seven suggests some modifications to the model, which make it appropriate for use by school boards in evaluating their own performance.

Chapter Eight offers recommendations for relating performance appraisal to administrative compensation. This chapter offers two models for developing a compensation plan that is related closely to the outcomes of the performance appraisal process. An example of the implementation of the appraisal and compensation programs with an administrator who must be rated as unsatisfactory is also included. The final chapter offers suggestions for developing and implementing a new administrative appraisal program within a school district. This chapter identifies the pitfalls to be avoided in dealing with this often emotionally charged subject and suggests strategies for creating a new appraisal program in a timely and sensitive manner.

Throughout the book the authors make the argument that the current unsophisticated state of administrative performance appraisal should not and cannot continue. Practical guidelines are provided that will enable superintendents to develop and implement appraisal programs that are both efficient and effective. Progress in this critical area will enable school districts to make a positive impact upon administrative performance and thus improve the education of our students.

CHAPTER ONE

The Context of School Administration in the 1990s

"THE Technomic School District will open two hours late today, and there will be no morning kindergarten." It's only 6:00 A.M. on a snowy morning but you, as a school superintendent, have already made a decision that will affect the safety of thousands, will disrupt the normal routines of countless households, and will be debated in faculty rooms and coffee shops throughout the school district.

As the snow swirled across your windshield on your early morning road check you might have reflected that the decision to announce a school delay because of bad weather offers a perfect model for the decision-making process of the school superintendency. You most often must make decisions based on partial information, expert opinion is often divided, and you will be second-guessed by members of your staff and community who are often wrong, but never in doubt. Such variables also are at play with the subject of this book, performance appraisal of the school district's management team.

Leadership in an enterprise as public as our schools is profoundly influenced by the social milieu in which the schools function. School officials, as well as politicians, clergymen, business leaders, and judges, are among a short list of community leaders that are common to everyone's experience. The manner in which these figures are perceived and portrayed in the popular culture powerfully influences the context in which school administrators must accomplish their tasks.

In this chapter we will briefly address broad cultural trends, as well as specific initiatives in the field of education, which together form the context for school administration in the 1990s. The magnitude of the challenges presented by the social and cultural environment in which school leaders must function requires that administrative performance appraisal be given a far higher priority in the decade ahead. The purpose of this book is to provide school executives with a mechanism to implement an appraisal program that will both nurture and develop the leadership talents of our school administrators.

Impact of the General Culture on School Leaders

The first cultural reality that school leaders must face is that, in America, the public views all leaders with a certain skepticism, if not distrust. This egalitarian, and perhaps anti-authoritarian habit of mind, was first noted early in the nineteenth century by Alexis de Tocqueville in his classic *Democracy in America.* Over the span of several generations, Americans have maintained this viewpoint to the extent that many would endorse Lord Acton's famous dictum that "power tends to corrupt and absolute power corrupts absolutely."

This healthy skepticism that Americans have toward their leaders has been more than vindicated by the public disgrace that has been visited upon so many sectors of our leadership class over the past few decades. From Watergate to Iran-Contra, from the S&L scandal to the federal budget debacle, our political leaders have proven to be both untrustworthy and inept.

The credibility of corporate leadership has been seriously compromised by the manipulations of Wall Street financiers, the junk bond fiasco, and the collapse of the financial empires of such publicly celebrated figures as Donald Trump. Even spiritual leaders as diverse as Roman Catholic prelates in Chicago and New Orleans and Fundamentalist Christian preachers have brought disgrace and discredit to their calling.

Measured against these major failures of leadership in the political, corporate, and religious arenas, the shortcomings of school leadership seem almost trivial. Nevertheless, the public's perception of school leaders has been and will continue to be influenced by the behavior and performance of leaders in all segments of society. Our own profession has been disgraced by periodic disclosures of corruption such as the case of the Florida superintendent who misappropriated school funds to purchase gold-plated bathroom fixtures for his own home. Thus in the decade of the 1990s, the trust and respect that school administrators would like to be granted by the public will not be freely given but will need to be earned by each individual administrator.

School Reform and the School Administrator

Successful school administration in the coming decade will require administrative personnel with a higher level of both technical and people skills than are commonly found among our current cadre of school administrators. The advent of site-based management and participatory decision making by teachers have important implications for building administrators, central office personnel, and even the school board.

Superintendents of local districts at the dawn of the 1990s are largely the products of graduate programs of the 1960s and 1970s. We began our administrative careers at a time when the paternalistic management practices common to the post World War II period were giving way to the democratic leadership theories of McGregor and others. Most of us who have recently held senior administrative positions can reflect on the early days of our teaching careers when our superintendents were clearly autocratic and at best were viewed as benevolent dictators. As teachers, we were likely to greet a summons to the superintendent's office with no greater enthusiasm than we felt as students when called to the principal's office. These new theories of democratic leadership produced an entire generation of administrators who were taught to involve teachers and other staff members in the decision-making process.

Many practicing administrators had barely assimilated the participatory management leadership style into their repertoire before the reform movement of the 1980s made even these modest steps toward shared governance seem dated and insufficient. Prestigious reports on school reform, prominent leaders of teacher unions, and management experts alike have universally proclaimed the desirability and even necessity of providing teachers with a significant voice in determining the programs and practices within the school and classroom.

Those of us who have spent entire careers attempting to cope with the adversarial environment that often accompanies collective bargaining are now being asked to assimilate a true paradigm shift. How able will we be to successfully integrate the roles of adversaries at the bargaining table with those of collaborators on school-site councils and other shared governance forums?

The concepts of shared decision making and site-based management are being vigorously applied on the factory floors of many of our large corporations. One of the most prominent examples of this new management thrust can be found in the General Motors' Saturn plant near Nashville, Tennessee. The traditional adversarial stance between union and management in the automobile industry is being replaced by a more collaborative relationship featuring teamwork and decision making by workers on the assembly line. The apparent success of these strategies in the industrial world will contribute to the impetus for similar changes in the school environment. We as educators, by drawing on the historical tradition of schools as communities of scholars, should be able to quickly move our schools beyond the industrial factory model toward the more collaborative approach now flourishing in the business world.

Not only are teachers to be more involved in the decision-making process, but the school itself is perceived as the unit for educational accountability. Decentralization of decision making is a recent trend in the business world, which is being urged upon the schools as part of the school restructuring movement. Many large corporations are downscaling their main office operations and allowing for a greater level of decision-making power at the district or regional office levels. The expectation is that this decentralization effort will have the twin benefits of better decision making coupled with a cutback in the size of the upper management bureaucracy. This decentralization thrust has particular applicability in the large school districts in our major cities.

The educational reform literature rarely takes these twin concepts of site-based management and teacher empowerment beyond the theoretical stage to examine the practical implications of such a radical departure from current practice. Clearly the building principal will need far more highly honed people skills and group process skills than are currently the norm among principals. Participatory decision making will also require administrators who truly believe that staff members in general have something positive to contribute to the leadership equation.

Such administrators will likewise need the capacity to reserve judgment, seek out and consider all points of view, and implement with vigor decisions with which they might not totally agree. Only individuals secure in their own person and convinced of the talents and good judgment of their staff colleagues will be able to thrive in this new environment.

The superintendent and other central office administrators will find that a shared governance mode may greatly complicate the delegation of assignments to a principal as well as the process of goal setting between the superintendent and the principal. Evaluating the extent to which goals and objectives have been achieved will also be more complex given the number of staff members involved in both formulating and implementing the goals and objectives.

As a young administrator, one of the authors worked for a superintendent who supervised his principals through occasional phone calls, brief meetings in the superintendent's office, and by reviewing monthly reports submitted by his principals. Not once in two years did this superintendent actually visit the school. Such "role models" need to be totally rejected by administrators of the 1990s.

School boards will need to become more adept at separating their policy-making function from their all too common proclivity for micro-managing the school district. Even union leadership may find threats to their power base as individual schools request waivers from certain provisions of the district union contract. In our experiences with union leaders we have seen them confronted with this issue of balancing central control with shared governance, just as it is faced daily in the superintendency.

A recurring decision in the authors' experiences that represents a virtual parody of a shared governance model involves the development of a school calendar for the ensuing school year. The relatively trivial issue of scheduling snow makeup days and the length of spring vacation seem to take on cosmic importance as teachers are surveyed about their preferences. Teacher and administrative committees then meet in an attempt to reach consensus on a proposed school calendar. This exercise has always struck the authors as a case where the time and energy required for this staff involvement effort is disproportionate to the benefits derived from the process.

School board members and teacher union leaders may, in fact, have the most difficulty adapting to the emerging power alignments. Virtually none of the teacher leaders and relatively few school board members will have either the management theory or experience to deal effectively with the subtle nuances of collaborative leadership.

The move toward greater educational accountability is also a major thrust of the school reform movement. In the 1970s writers such as Leon Lessinger in his *Every Kid a Winner* developed elaborate techniques to monitor student achievement from a district-wide perspective as a means of providing accountability. Such programs were never widely adopted, so that to the present day there is a continued call for better accountability.

As educational leaders, we have not distinguished ourselves by our almost universal aversion to being held accountable for the achievement level of our students. We have tacitly or openly supported the claims by our teachers and their associations that educational goals are too complex and broad to be adequately addressed by standardized tests. We have also been quick to agree that teachers should not be held accountable for the performance of students whose homes and communities do not provide support and encouragement for education. Rather than to seek proactively to develop better student accountability measures, we have been content to criticize the deficiencies of the existing measures.

While many states have mandated the use of some form of standardized testing to measure the performance of school districts, even these traditional accountability measures are being challenged as inadequate, unreliable, and invalid. Some states are experimenting with or requiring writing samples as a component of the traditional multiple choice assessment tools. Portfolio testing is yet another proposal whereby samples of student work gathered over time would become part of the formal assessment process.

The daunting challenge for today's superintendent is to provide this higher level of accountability in an era of school-based management and shared governance with teachers and other professionals. It will be interesting to observe the accountability measures that might be proposed by a school faculty after it has defined its mission as a school.

During a lifetime of experience with school faculties in developing long-range plans and conducting evaluations through regional accrediting agencies, the authors have yet to encounter a faculty anxious to develop accountability measures that would utilize hard data to evaluate the performance of the school and its faculty. It will require superior leadership ability on the part of the superintendent to persuade site-based decision makers to identify and adopt meaningful accountability measures.

Schools of Choice is a variation of the voucher concept that could conceivably affect school administration in the 1990s. Should this concept be widely implemented, a highly debatable proposition, schools would of necessity become market driven and market sensitive. Some of the marketing techniques currently employed by commercial tutoring schools and even diet centers could be practiced by public schools seeking to recruit and retain students.

The underlying assumption of the Choice movement is that parents and students will seek to attend certain schools because of specific academic advantages that these schools might offer. The authors worked together in a school district almost twenty years ago, as superintendent and high school principal, in which the two high schools in the district offered distinctly different instructional programs. Students were allowed to choose either school. Although students sometimes sought a transfer to the other high school for reasons of athletic participation, personal convenience, or family tradition, there were few instances where transfers were requested for academic reasons.

Early results of the Choice experiment in some states indicate that fewer than one-half of one percent of students are electing to attend school outside of their normal school attendance areas. Furthermore, a majority of those who are choosing another school are doing so to accommodate the work or transportation convenience of parents or because of the athletic interests of the student.

Schools of Choice would give rise to a new breed of school administrators who would of necessity have a flair for public relations and entrepreneurship. The performance of such administrators would be measured in the coin of the business world, the bottom line. Even should Schools of Choice fail to materialize because of practical problems with the concept, some of the accountability notions inherent in the idea will affect the expectations placed upon public school administrators. School administrators can expect to have the performance of their students publicly compared to the performance of students in other schools to an even greater extent than has been common in the recent past.

Wide acceptance of the Choice concept could result in the abandonment of the public schools by the middle class. Rather than fight the political battles of financing a quality program at their local public school, many people find it more expedient to simply remove their students from a setting where the needs of all students must be addressed. In this scenario, the existing level of political and financial support for the public schools would be further undermined, and our public schools would once again become pauper schools. Should the middle class leave the public schools, both financial support and educational achievement would decline significantly.

The trends toward shared governance and greater accountability will create special leadership tensions for the administrator of the 1990s. Teachers generally resist and do not personally value quantifiable accountability measures favored by the general public. Teachers prefer to measure school quality in terms of input measures such as class size, salary levels, and size of instructional budgets.

The public increasingly judges school quality in terms of output measures such as achievement test scores, college attendance rates, and dropout rates. Convincing teachers to measure their success in terms acceptable to the public will require administrative leadership skills of great sophistication. The challenge for the administrator will be even greater since decisions about the goals of the school will be made in a more collegial style consistent with the shared governance concept.

Supply and Demand of School Administrators

This brief discussion of the school leadership challenges of the next decade raises the interesting question of supply and demand of school administrators. Where will we get the highly skilled and dedicated professionals to lead our schools through the 1990s? Can we rely on a steady supply of quality school leaders under current practices and conditions? Unfortunately, there are no easy answers to these questions. We are going to have to create more effective performance appraisal and administrative development programs to increase the effectiveness of those that can be recruited to these positions, regardless of their initial preparation and aptitude for the job.

We believe, for the reasons discussed below, that the supply of quality administrative candidates in the next decade will be extremely limited and that the quality of appraisal and development programs will have to compensate for the deficiencies in the size and quality of the talent pool. It will become a priority task of the local superintendent to initiate and implement a far more sophisticated and effective performance appraisal system than has previously existed in public school administration.

Demographic trends in public education over the past twenty years will lead to a shortage of administrative candidates in the 1990s. During the late 1970s and early 1980s there were fewer new teaching positions available, due to a declining birthrate followed promptly by declining school enrollments. Not only were school districts failing to hire new teachers, but in many cases they were furloughing their relatively young teachers.

Both authors, during their tenures as school superintendents, were involved in the painful process of recommending and implementing teacher layoffs in response to precipitous declines in student enrollments. In each case, legal constraints relating to the teacher furlough process required that the younger, and often more effective, teachers were the ones to be furloughed. Thus the number of current public school teachers in the thirty- to forty-year-old range is disproportionately small because of the comparatively few teaching positions available when these teachers were entering the profession.

The thirty to forty age cohort is the group that supplies the candidates for entry-level administrative positions. Thus the talent pool for new administrators will be especially small just at a time when the demand for new administrators will be especially large due to retirements and a resurgence in student enrollments. These supply and demand constraints provide little reason to expect that the overall quality of administrative candidates will improve or even maintain the current level of quality over the next decade.

A second factor mitigating against a sufficient supply of quality administrative candidates is the lack of a reasonable salary differential between what a teacher would earn in an entry-level administrative position versus continuing as a classroom teacher. This situation has long been a problem in school administration, but recent changes in the structure of teacher salary scales will soon lead to a crisis situation.

As young heads of households in the 1960s and 1970s, the authors were able to view school administration not only as a challenging career, but also as a means to provide their traditional families with a significantly higher standard of living than would be possible on a teacher's salary during that time. Today's potential administrator is almost certainly part of a two-income family. The relatively small increase in salary that the candidate would earn as a school administrator would have only a marginally positive impact on total family income.

The general increase in teacher pay over the past ten years is the principal reason for the diminishing differential between teacher pay and administrative pay. A second phenomenon, salary schedule compaction, is also significantly affecting the differential. Salary schedule compaction is the process whereby fewer years are required for a teacher to reach maximum salary. In the past it often took a teacher twenty years or more to reach maximum salary. Increasingly teachers are negotiating contracts where maximum salary can be reached in twelve to fifteen years.

This is precisely the point in their careers when teachers are most likely to be ready to make the switch to administration. Such teachers are now finding that their salary as a twelve-month administrator would be no greater than that of a nine-month teacher. Many potential administrators are concluding that the extra stress and responsibility of school administration is simply not reasonable given the lack of a financial incentive.

The trend toward encouraging teachers to become more involved in the decision-making process, staff development programs, and inducting new teachers to the profession all have an effect on the recruitment of administrative personnel. Many teachers will now be able to exercise many of their leadership skills and desires without leaving the classroom. They will also be able to have a greater direct impact on decisions and school operations as teachers than had previously been possible.

While engaging in these activities they will often be eligible for released time, extra compensation, or both. As administrators they would typically be expected to assume extra responsibilities, with neither additional compensation nor a decrease in their normal responsibilities. It is understandable that a lead teacher assignment might be more appealing to a teacher than becoming an assistant principal in charge of discipline at a secondary school.

One of the authors has for fifteen years been superintendent in a school district with 150 teachers, in which every attempt has been made to professionalize teaching and maximize opportunities for teachers to have meaningful involvement in the decision-making process. During this entire period of time not a single teacher from this district has entered school administration in this district or in any other school district.

Changes in lifestyles and family structure are also having an impact on school administrators. The two-income family is as common for the school administrator as it is in the culture at large. The problem of finding a job for the trailing spouse is limiting the mobility of many administrators and thus decreasing the number of applicants for a given position. Just as corporations are finding that their executives are less willing to relocate, school districts are finding fewer candidates that are willing to move to a new area in search of an administrative position.

Contrast this situation with the world of the 1960s, when the authors began their careers in school administration. Almost all administrators were males, with wives and families ready, if not always anxious, to follow the breadwinner to new opportunities for career advancement. Not only was the wife expected to graciously accept frequent moves in her role as trailing spouse, but often she was also expected to prominently involve herself in the social and civic life of the community.

A household with children in which both parents hold full-time jobs creates an additional dilemma for the would-be administrator. Is it justifiable to take a job with the time and energy commitments required of school administration when one must also carry a major responsibility for raising the children? A number of potentially excellent administrators are facing this question and deciding that a classroom teacher is better able to meet family responsibilities than is a school administrator. These well-motivated and reasonable individual decisions by countless potential administrators are further limiting the supply of future administrative candidates.

Throughout this chapter we have discussed the many ways in which trends in the general culture, as well as initiatives specific to the field of education, will both challenge and constrain school administrators during the next decade. Within this context, the school administrator will be expected to produce better educational results while working with a school clientele suffering from an increasing array of social, economic, and cultural deficiencies. We are convinced that the performance appraisal models provided in this book can better equip our school leaders to meet and conquer these formidable challenges.

Suggested Readings

Baldwin, G. H. 1990. "Collective Negotiations and School Site Management," *West's Education Law Reporter,* 58(4):1075-83.

Etheridge, R. E. 1989. "A Sensible Approach to Solving Problems in Schools," *NASSP Bulletin,* 73(518):38-39.

Honeyman, D. S. and R. Jensen. 1988. "School-Site Budgeting," *School Business Affairs,* 54(2):12-14.

Karant, V. 1989. "Supervision in an Age of Teacher Empowerment," *Educational Leadership,* 46(8):27-29.

Lagana, J. F. 1989. "Managing Change and School Improvement Effectively," *NASSP Bulletin,* 73(518):52-55.

Langlois, D. E. 1989. "Today's School Leaders Don't Measure Up to the Giants of Old," *The Executive Educator* 11(4):24-26.

Lindelow, J. and S. Bentley. 1989. "Team Management," in *School Leadership: Handbook for Excellence.* Eugene, Oreg.: Eric Clearinghouse on Educational Management, 19 p.

Luby, G. 1983. "Shared Governance and Instructional Improvement in an Urban School District," *NASSP Bulletin,* 67(466):12-17.

Phillips, P. R. 1989. "Shared Decision-making in an Age of Reform," *Updating School Board Policies,* 20(3):1-4.

Pulling, J. 1989. "We Can Learn from Business—And Teach a Thing or Two," *Executive Educator,* 11(4):38-39.

Raelin, J. 1989. "How to Give Your Teachers Autonomy Without Losing Control," *Executive Educator,* 11(2):1920.

Regan, H. 1988. "Up the Pyramid: New Leadership Opportunities for Principals," *NASSP Bulletin,* 72(506):84-86.

Sheive, L. T. 1988. "New Roles for Administrators," *Educational Leadership,* 46(3):53-55.

Tewel, K. 1989. "Collaborative Supervision—Theory into Practice," *NASSP Bulletin,* 73(516):74-83.

Tursman, C. 1989. "Ways to Fight Teacher Burnout," *School Administrator,* 46(3):30, 35.

Wynn, R. and C. Guditus. 1984. *Team Management: Leadership by Consensus.* Columbus, Ohio: Charles E. Merrill Publishing Co.

CHAPTER TWO

Performance Appraisal Priorities

THE brief review of the special challenges facing school administrators presented in the last chapter points to the need for more effective performance appraisal programs that specifically address the new challenges for the 1990s. In this chapter we will outline those elements of an effective appraisal program that should receive special attention during the next decade.

Involving Staff Members in the Appraisal Process

Interaction with staff has always been critical to the success of school administrators. This factor will assume even greater importance in an age of shared governance and the assumption of leadership roles by members of the teaching staff. The ability of an administrator to both motivate and coordinate staff activities, while at the same time fostering the development of leadership skills in subordinates, will require a high degree of open communication between the supervisor and the supervisee.

This open communication should extend to the performance appraisal process itself. In the paragraphs below we will specifically explore the concept of formally involving teachers in the appraisal of the principal. The same general approach could be followed for including subordinates in the evaluation of all levels of the administrative team.

Anyone who has served as a school administrator realizes that he or she is constantly being evaluated, not only by subordinates, but by students, parents, and the general community. These evaluations take place in the faculty room, in the student cafeteria, and at the checkout counter of the local supermarket. Thus, the question is not whether subordinates should evaluate the administrator, but rather, whether there will be some formal mechanism for the administrator to become aware of the opinions of his or her subordinates.

Involving teachers in the process of evaluating principals is not common practice today. Approximately 86 percent of U.S. school systems have formal procedures for evaluating their school executives — but only 14 percent ask teachers to evaluate principals. Why is it that this approach has not been tried on a wider basis? Is it because school executives lack self-confidence? Perhaps it's because teachers can't be trusted. Or could it be that administrators have a reputation for being vindictive — punishing teachers who might give them negative evaluations? We don't know why only a small percentage of school systems include teachers in the evaluation process, but — speaking from our observations of several school systems that do — we know the rewards can be great.

Under one of the plans we have seen, school administrators evaluate themselves, are evaluated by their staff members, and are evaluated by their immediate superiors. The evaluations cover four major areas of administrative responsibility: planning and evaluation, administering, making decisions, and communicating. Included within these four areas are nineteen general job-related tasks and sixty-one specific criteria. All these areas are incorporated into the instrument that supervisors use in evaluating principals. Teachers use a shorter, twenty-five-point form based on the longer evaluation instrument to assess a principal's performance.

Teachers rank their principals on a scale from 1 to 5 for each of the twenty-five items. Under "planning and evaluating," for example, a teacher would give a principal the score of 5 if he or she always "formulates appropriate school goals and objectives" — and a score of 1 if he or she rarely does so.

This type of evaluation plan has met with widespread approval. Teachers appreciate the plan because it allows them to know what the school system expects of administrators — and it allows them to participate in the development of their principals' management-by-objectives plans. One teacher who has participated in such evaluations for several years points out that his principal has worked hard to overcome the deficiencies teachers called to her attention — and adds that teacher/principal relations have improved considerably since teachers started giving principals feedback. Other teachers point out that morale has improved in their schools since teachers have been asked to evaluate their principals and supervisors.

Principals, understandably, might feel threatened by the plan at first. Said one principal, "I thought (the evaluation plan) was going to be a popularity contest." The same principal, however, says now he's more open with staff members. He adds, "They give me good, helpful feedback. It hurts at times, but it has improved the overall climate in our school."

Another principal we know happily shares the results of these teacher evaluations of his performance with the superintendent during performance appraisal conferences. The results of these surveys of teachers' opinions of their principal are so positive that they serve as documentation to the superintendent that the principal is doing an outstanding job in the areas of staff relations and communications.

In the case of the principalship, the following procedure could be used for initiating a process for involving teachers in the principal appraisal process:

(1) At a faculty meeting, the principal would explain the importance of teachers taking part in the evaluation of the principal. He or she would note, for example, that such a process would help the principal improve his or her performance and would promote open communication between principal and staff.

(2) The principal would give specific directions for completing the form and would leave the room while the teachers are completing the form.

(3) A trusted teacher would collect the completed forms. This teacher then compiles a composite report with an average score for each of the twenty-five criteria—and destroys the individual forms before giving the composite report to the principal.

(4) The principal uses the results for his or her information only. Results are not shared with the principal's supervisor.

After working with hundreds of school executives and thousands of teachers in several school systems, we are convinced that many administrators want to be evaluated. They want, too, to be part of the team that develops the criteria for assessment, and they would welcome the involvement of teachers.

The need for inviting teachers to evaluate principals is based on the following reality: no one is in a better position than teachers to determine the performance level of the principal. Teachers see their principals in action every day, and they know more about principals than administrators sometimes realize. Involving teachers in the evaluation process will provide results that will be both pleasing and surprising.

The practice of subordinates formally evaluating their supervisors can be utilized at all levels of the administrative hierarchy. Building and central office administrators could have formal input into the evaluation of the superintendent, and the superintendent could play a role in the process that the school board uses to evaluate its own performance.

The concept behind the recommendation to involve staff in the evaluation process is that judgments regarding the effectiveness of administrative personnel are among the most important and significant decisions made within an organization. Surely the concept of shared governance and staff involvement in decision making implies a meaningful role for all staff members in the performance appraisal process.

The Coaching Component in Performance Appraisal

A second priority in performance appraisal programs for the 1990s relates to the concept of coaching. The supply and demand constraints on new administrative talent reviewed in the previous chapter point to the critical necessity to develop the talents and skills of those who will be appointed to administrative positions during the new decade.

A superintendent has many opportunities to play the role of coach. During a single two-day period in his superintendency, one of the authors received requests from two relatively new principals for help with the following dilemmas:

(1) A high school teacher, through a misjudgment, allowed students to show a Madonna videotape in class which had pre-

viously been banned for viewing by MTV as being too sexually explicit and suggestive. Several parents were outraged and one called the local paper to report the story. How should the principal react to the teacher, the parents, the students, and the news media?

(2) A parent of an elementary school student called the principal to demand that his child be assigned to another teacher because he believes that the recent interracial marriage involving his son's current teacher is immoral and scandalous. He threatened to come to the next school board meeting and read verses from the Bible that would serve to condemn the immoral lifestyle of this teacher. He believes that it is his religious duty to bring this matter to public attention. The principal wants some suggestions for responding to this parent.

(3) A high school student set fire to a poster on the bulletin board in the main hall of the school. The possible disciplinary actions available to the principal range from a ten-day suspension to a recommendation for expulsion. The principal is faced with the all-too-familiar problem of attempting to temper justice with mercy. The principal wants to discuss possible options with the superintendent.

(4) High school students enrolled in a course jointly sponsored by the Industrial Arts and Business departments wanted to produce T-shirts for sale containing strong political statements about hostilities in the Persian Gulf area. The principal denied their request, consistent with school policy forbidding the promotion of partisan causes by the school district. The students then complained of censorship and failed to see the distinction between the expression of personal beliefs and the possible perception that the school as a public institution could be viewed as promoting partisan political causes. The principal wants the help of the superintendent in developing strategies for dealing effectively with this volatile issue.

In addition to providing the benefit of his experience to his subordinates, a superintendent can learn much about the leadership abilities of his administrators during these coaching opportunities. Does the administrator gather sufficient facts before considering a course of action? Can the administrator lay aside personal emotional considerations in choosing a course of action? Is the administrator able to distinguish between issues that truly require the attention of the superintendent and those that should be resolved without central office consultation? These coaching opportunities will provide the superintendent with critical information about subordinates which can later be incorporated into the formal performance appraisal process.

The coaching component of performance appraisal should be an element in the evaluation of all segments of the administrative team hierarchy. In evaluating its own performance, the school board should appraise the extent to which it provides for the proper orientation and continued development of its own members.

In considering the area of relationship with its superintendent, the school board should evaluate the extent to which the board clearly states its expectations of the superintendent, evaluates the performance of the superintendent on a regular basis, and provides the superintendent with appropriate opportunities for professional growth. The school board appraisal system must clearly reflect the importance of the school board's coaching of itself and its superintendent.

Both authors are aware of cases where school boards have poorly handled their responsibilities to coach and evaluate their superintendent. Some school boards have seriously miscommunicated their true judgment regarding the performance of their superintendent. This failing has on several occasions reflected itself in the public spectacle of a school board renewing a superintendent's employment contract, only to dismiss him or her or buy out the contract within the space of a few months. The classic story of this type involves the superintendent in a New England school district who had his contract renewed at a public meeting only to be fired by his school board several hours later during an executive session.

The appraisal program for the superintendent should also give prominence to his or her role as coach for immediate subordinates. The role of the superintendent as coach can be played in a myriad of ways. The following example would be applicable to a superintendent who is new to a school district.

Every superintendent who starts a new job fantasizes at some point or another: "Wouldn't it be great if I could build an administrative team from scratch—filling every slot with the person of my choosing? That way, I could get right to work without having to waste time fretting over one or two weak administrators my predecessor has left behind."

But the real world is far from fantasy, and few superintendents have the luxury of handpicking a staff. In our combined thirty-five years as superintendents, we've inherited staff members whose performance ranged from marginal to outstanding. Sometimes, we're happy to say, the mediocre staffers turned into stars — but not without some help in the form of coaching. Superintendents have a duty to provide that help so that marginal performers can be transformed into valuable, productive contributors to the school system.

While a superintendent can never hope to transform everyone, he or she can reduce, eliminate, and prevent many types of nonfunctional administrative behavior. As a good coach, the new superintendent would be slow to form negative judgments of subordinates. Accurate judgments of individual performance cannot be made until the superintendent understands the school district as it existed before his or her arrival and as it currently functions. This takes time.

The new leader needs to understand how the system supports those who make a poor first impression as well as those whom the new superintendent has been warned about. In short, one can't assume too much during those early days. Pieces of a picture, however true they might be, do not provide a complete portrait. Making premature decisions about restaffing might be unwise and could rob the new superintendent of the support he or she might need for future decisions.

The new superintendent must also remember that judgments about administrator performance are relative. One person's turkey is another person's swan. The behavior of people within the system is a function of the values, norms, structures, and agendas of the school district as well as the specific expectations placed upon a given administrator. The new superintendent needs to view all staff members, including perceived poor performers, in terms of the current organizational system as well as the system that the board, community, and superintendent wish to create. This means determining what skills are needed and who has them.

Staffing schools according to an organization chart, for example, is fine on paper, but often proves impractical in practice. It makes sense, from our experience, to identify what the school system needs and then match those needs with existing talents. The resulting organizational chart might look convoluted, but people are more likely to be assigned where they are most needed and most productive.

We also have learned that staff members are less resistant to job transfers if the supervisor can offer a believable rationale based on the skills and abilities of the staff member. Transfers are also better received if staff members help to make the decisions. A significant number of people are affected by transfers, and it is standard practice to transfer people periodically.

These principles were closely followed by one of the authors during his tenure as superintendent of a relatively large school district. Within a short period of time he transferred many principals among the many elementary and middle schools in the district. There was a good deal of anxiety and even opposition among some of the principals and among many parents at the elementary school level. Due to detailed preliminary planning, solid rationales for the transfers, and the large number of administrators affected, the transfers were ultimately successful. Most of the administrators attacked their new assignments with renewed enthusiasm and infused their new schools with a new mission and purpose.

To accumulate all of this background information about the system and its employees requires time. And during this time the superintendent should be building the trust required for people to understand and accept, if not contribute to, the leader's vision of what the school system should be. Building a team of supporters is as important as building the trust of individuals. Marginal performers, for example, sometimes improve significantly when the rest of the team endorses the new organizational plan.

One way to learn about your organization as it is and might be—and to build trust at the same time—is to listen to and talk with your staff members at length. Simple as it sounds, many new superintendents ignore this practice. It's true that superintendents are gifted talkers. But it's a rare one who listens, who creates opportunities for weak administrators (and others) to discuss their frustrations, careers, aspirations, experiences, and interests. Yet knowing these things about staff members pays dividends when making job assignments—assignments that won't be resisted. Learning to listen—without encouraging staff members to complain and without making promises that can't be kept—is difficult but possible. This ability to listen is a major function of the superintendent as coach.

A good coach has a definite game plan that is clearly communicated to team members. While taking time to listen and learn, the superintendent must be aware of what he or she is communicating — and to whom. Every action, memorandum, and meeting should be viewed as a subtle yet explicit definition of the superintendent's agenda for the school district.

The new superintendent must be clear about what he or she expects and what he or she stands for. This gives all staff members, but especially the marginal performers, some early sign about new expectation levels. The more consistent the words and actions, the stronger the message. Remember, too, that most staffers — even poor performers — expect to adjust to and want to please the new boss. Clear signals from the superintendent help them to achieve these goals. Some long-term mediocrities might use the arrival of the new superintendent as an opportunity to turn over a new leaf. As coach, the superintendent will make sure that they have the clues they need to turn in the right direction.

Finally, the new superintendent needs to do some soul searching. What preconceived notions about people is he or she bringing into the new position? Is the superintendent starting the job with insecurity, suspicion, cynicism, or mistrust? Does he or she believe the best about people until they prove otherwise? A positive approach cannot of itself transform a weak administrator, but it helps the superintendent see the salvageable characteristics of poor performers.

The concept of proper supervision of staff members is of obvious importance and yet is often honored in the breach. By this we mean staff members are not coached — i.e., reminded that they play for the same team and that the superintendent knows their short-term and long-term objectives, as well as their work styles, problems, and achievements. Supervision means holding periodic meetings, during which the superintendent listens, teaches, prods, suggests, paves the way, corrects, and applauds successes. In short, effective supervision means setting clear expectations, staying in touch with what subordinates are doing, and providing feedback to them regarding their efforts.

It's especially easy for the superintendent not to properly supervise marginal performers because they are often adept at avoiding supervisory efforts and clever at making the supervisor believe that he or she is the problem. Interacting with these people is usually unpleasant, yet they require the superintendent's best supervisory efforts. Such employees often improve once they realize that the superintendent is not going to go away and that the supervisor will insist that his or her expectations be met.

Marginal performers are survivors. They've learned that if they can make it through the semester, the end-of-year conference, or one more evaluation cycle, the pressure will be off for another year or so. Good supervision keeps appropriate pressure on all the time. Strange as it seems, most poor performers don't experience that pressure because it's easier for the supervisor to bury the problem. A good coach puts the pressure on the weaker members of the team — where it belongs.

Good coaching also means knowing when to increase, reduce, or change someone's job responsibilities. Many marginal performers started out as effective staff members, but simply died on the job — usually from lack of supervision. Assistant principals are especially vulnerable to this problem. Seven years, we believe, is the maximum length of time anyone should hold the same job in a school system. After that, administrators slowly lose the ability to be effective.

We have found that the professional growth of the assistant principal can be promoted by involvement with district-wide committees and/or responsibility for certain district-level projects. It is also important that assistant principals be given specific responsibilities for teacher supervision and evaluation and for involvement in the curriculum review and development process.

Supervision means helping people stay professionally awake. Administrators need to be encouraged to take advantage of worthwhile graduate programs, workshops, and conferences — and of other professionals who can help them grow beyond the narrow focus of their daily routines. Staff members should be assigned to district-wide projects, such as leading staff development programs, conducting research, and evaluating special projects.

In this process staff members grow, jobs get done (usually well), and the superintendent learns more about the skills of his or her people. Small school districts or districts with lean central office staffs are perfect places to apply this strategy.

In summary, the superintendency demands working with and through others. It's easy to work with those who are willing. It's much more difficult, yet potentially more rewarding, to work with the hard-to-reach. By keeping one eye on the evolving needs of the district and the other on staff members' individual needs and strengths, a superintendent can impact significantly on administrative performance.

Coaching at the Building Level

The coaching concept can also be applied to the supervisory relationship between the principal and his or her teachers. This approach seems especially suited to the collegial and shared governance mode of school operations that will become more common in the 1990s.

The performance appraisal program for principals must value and reward a rigorous program of teacher supervision and evaluation. The appraisal program should highlight the principal's administrative skills as an instructional leader and supervisor and should address coaching as a technique for improving teacher performance.

The major characteristics of an effective staff evaluation program to be implemented by principals are relatively few in number. The starting point for such a program should be thoughtful and detailed analyses of teaching practices. Such analyses would help to identify unsatisfactory teachers, who could then either be helped to improve or removed from the classroom. Marginal teachers could be identified and offered appropriate help. The majority of good and competent teachers would appreciate evaluations that reinforce their successes, point out their weaknesses, and offer suggestions for expanding their repertoire of teaching strategies.

And the excellent teacher? Even in a well-conducted system of teacher supervision and evaluation, it might be easy to fail to sufficiently recognize the truly excellent teacher. Why bother telling someone he is excellent when he probably knows it already? But what an injustice it is to deny such a person praise for being what we wish all teachers were. And how foolish to miss the chance to reinforce, at a conscious level, what might merely be instinctively good teaching techniques. A good coach extols the achievements of the outstanding performer as a role model to be emulated by others.

The concept of coaching as an integral part of the supervision process will take time to implement. How are we to pursue this ideal amid realities that have already doomed many fine proposals? We're talking about the frantic busyness of the central office, the demands of superiors, the ire of parents, the bravado of students, the complaints (or, worse yet, the complacency) of teachers, and the grumbling of unions.

An incident early in the career of one of the authors clearly illustrates the perennial problem of time pressures on administrators and the search for a workable solution. At a district meeting for administrators, a central office supervisor was explaining to building principals his plan for them to keep a detailed time log of all of their activities for one week. This information could then be analyzed in the hope that individual administrators would be able to spend their time more efficiently.

This project was greeted with some skepticism by several of the principals in attendance. They viewed the assignment to keep a time log as but another illustration of the unreasonable demands on their time. One principal, in particular, was infuriated at this further drain on his limited time resources. Finally, in exasperation, he pounded his fist on the table and blurted out, "Damn it, I don't have time to figure out what I'm doing."

Granted that community relations, plant management, personnel decisions, student discipline, curriculum matters, and financial management all are important concerns. These aspects of school management exist, however, only because parents send their children to school to learn. Important though these matters might be, top priority surely belongs to good teaching. The principal needs to be encouraged to delegate whenever possible. Secretaries can handle routine tasks, other properly trained people can deal with small crises, and classroom teachers often can cope with the minor discipline problems they send to the principal's office. Encouraging staff members to take responsibility for their own problems will be good for them and will have a positive impact on the principal's schedule.

The superintendent, the instructional leader of the school system, needs to facilitate better supervision on the part of his or her subordinates. Three concrete recommendations follow:

(1) Institute a team approach to evaluation, using all of the supervisors available. These would include assistant principals, department heads, and central office personnel—all in addition to the principals.

(2) Put teacher supervision and evaluation at (or close to) the top of a principal's many tasks. Rearranging priorities means rearranging the order in which things get done. Tasks that were close to the top of a principal's list may now be close to the bottom.

(3) Help to conserve principals' time for supervision by conducting fewer and shorter meetings, carefully screening work before sending it on to them, requiring fewer reports, and communicating by telephone.

The type of supervision through coaching that we are advocating also implies a high level of skill on the part of the principal. Many practicing administrators are unfamiliar with the research on effective schools and effective teaching. Investing some time and money resources to train administrators in the art of classroom observation and conferencing of teachers will produce significant dividends. Such a training program should include all administrators and should be mandatory even in the face of inevitable resistance from some principals.

When the opportunity arises to recruit new administrators, superintendents should consider adding a step to the process that would provide a measure of the candidate's skills as a supervisor. Have each candidate view a videotape of a lesson, write up an evaluation report, and conduct a conference with a staff member playing the role of the classroom teacher. Although many candidates can identify supervisory skills in an interview, this technique will allow administrative candidates to demonstrate their skills.

All of the above suggestions and techniques are designed to help the principal to properly supervise and evaluate teachers. The coaching component has been woven through this discussion as an approach particularly suited to the staff relations climate of the 1990s. Improving supervision and evaluation through coaching will improve teaching performance and thus the quality of education for students.

One final comment should also be made with regard to the supervision and evaluation of teachers. Much has been made by principals concerning the problems that accompany their dual roles. They complain of the impossibility, or at least the difficulty, of creating a supportive, trusting, and collegial relationship with teachers throughout the year, only to "betray" that role with either an unsatisfactory post-observation report or end-of-year summative evaluation. We do not agree with this perceived role dilemma. Inherent in any supervision of a teacher are judgmental prerequisites. Before principals can support a teacher, they must evaluate to determine areas of weakness and to develop remediation strategies. We see supervision and evaluation activities as complementary, not mutually exclusive. One performs both activities simultaneously, though the proportion of each activity shifts from situation to situation.

Relating Accountability Measures to Performance Appraisal

A third priority for performance appraisal in the 1990s relates to the need for more quantitative accountability measures to be built into the evaluation process. In the past, far too much emphasis has been placed upon process variables in evaluating the performance of administrative personnel. This attention to administrative behaviors and procedures is a necessary but not sufficient component of the evaluation process.

The computer has provided us with ready access to a wealth of data relating to student achievement, student attendance, dropout information, budgetary data, and similar information that was previously unavailable or difficult to retrieve. Current computer programs operating in most public school systems are able to manipulate these data in a myriad of ways that make them useful for evaluating the success of programs and the administrators responsible for them. Software programs for both student records and attendance are available for freestanding PCs and networking systems for IBM, Apple, and other major computer companies.

The use of computer-generated data as an accountability measure can best be incorporated into a management-by-objectives performance appraisal system. The superintendent and principal might agree, for example, that student attendance rates need to be improved at the high school. The principal would be asked to formulate an action plan to reach this goal that would include specific initiatives to bring about the desired objective.

Both parties would agree upon the particular data that would be gathered to evaluate the extent to which the objective is achieved. Data would be gathered and jointly discussed several times during the year so that needed modifications in the attendance improvement projects could be made. At the end of the year cumulative attendance data would be used to evaluate the progress, or lack thereof, that had been realized on this particular objective. Since the superintendent would have worked closely with the principal throughout the year, he or she would be able to determine accurately the extent to which the principal had met the objective.

As a practicing district superintendent, one of the authors asked every principal in his district to prepare an objective for improving student attendance in his or her building. By closely monitoring attendance data for the year and meeting periodically with each administrator to discuss the attendance objective, measurable improvements in student attendance were registered at each school. At one school this initial effort served as a baseline for a multi-year objective that led to steadily improving student attendance at that school over a period of several years.

Similar types of data gathering could be used to quantitatively evaluate a principal's skills in the area of teacher evaluation. As a beginning, districts could adopt and enforce policies that (1) discourage principals from inflating evaluations or postponing dealing with incompetent or marginal teachers; (2) discourage principals from avoiding the issue by passing the poor performer on to someone else in the district; and, on a more positive note, (3) encourage principals to provide real instructional leadership.

To help avoid rating inflation, a school district might institute exit interviews with parents leaving the district to ascertain whom they judge to be particularly outstanding or poor teachers. Teacher observation reports and evaluations should be reviewed by the superintendent, assistant superintendent, or curriculum specialist. One approach to a systematic review of observations is a computer-based tracking system that could manage "housekeeping functions" and track the quality of both the teaching and the evaluation process.

Using a computer, one could record and then retrieve a great deal of valuable information. Such pieces of information might include the total number of observations made by a principal, the time of day and dates on which observations occur, the length of the observation and type of lesson observed, and the distribution of ratings given to teachers by each administrator.

Accountability by the principal for the proper evaluation and supervision of teachers can be further emphasized by directly relating a performance salary program to the staff evaluation function. Providing proper instructional leadership to teachers should rank as one of the major determinants of the principal's evaluation and thus salary increases.

The accountability measures employed as part of the evaluation process should be of a type that could easily be used to gauge the progress of a school and a faculty. These performance standards could then be shared with the staff and the community at large. The principal is in a position to constantly hold these standards before the staff and students, so that all interested parties are accustomed to holding themselves accountable for the achievement of specific, measurable goals of the organization.

Streamlining the Process

The last priority for an effective performance appraisal system for the 1990s is ease of use. A review of the literature on performance appraisal demonstrates that over the years many elaborate and worthwhile administrative evaluation programs have been developed or proposed. The reality in the field, however, has been to conduct the performance appraisal process in a perfunctory manner, if at all.

The goal in performance appraisal is to develop a program that is both valid and reliable. While the program should be thorough, the process should not be cumbersome or complicated. The program should be easily understood by the supervisee, documentation should be straightforward and easy to utilize, and final reports should be succinct and definitive.

The appraisal program should depend heavily upon one-on-one communication between the supervisor and the supervisee. Such discussions should be held frequently and should be viewed as formative coaching sessions rather than as summative evaluation sessions. Such periodic coaching sessions should be held at the work site of the supervisee to the extent that this is possible. This is particularly important in appraising the work of building administrators.

Completing the process through a summative evaluation conference and completion of required documentation should be quick and easy. The person with the greatest influence on the effectiveness of the performance appraisal process is the superintendent. The first duty of all superintendents is to set a high standard for the evaluation process through the example of their own commitment as they appraise the performance of their immediate subordinates. They must also encourage and even insist that the school board play its proper role in evaluating the superintendent and the school board itself.

In the chapters that follow, the concepts introduced in this discussion (staff involvement, coaching, accountability measures, and ease of use) will be featured prominently in the recommended appraisal programs. Special attention to these factors will contribute to the development of a performance appraisal system well suited to the 1990s.

Suggested Readings

Asbaugh, C. R. and K. L. Kasten. 1987. "Should Teachers Be Involved in Teacher Appraisal?" *NASSP Bulletin,* 71(500):50–53.

Barnes, R. E. 1987. "Help Teachers Help Themselves," *Executive Educator,* 9(9):23, 39.

Bottoni, W. R. 1984. "How to Evaluate and Improve the Principal's Performance," Paper presented to National School Boards Assocation. 9 pages (ERIC Document ED247644).

Grimmett, P. P. 1987. "The Role of District Supervisors in the Implementation of Peer Coaching," *Journal of Curriculum and Supervision,* 3(1):3–28.

Hopfengardner, J. D. and R. Walker. 1984. "Collegial Support: An Alternative to Principal-Led Supervision of Instruction," *NASSP Bulletin,* 58(471):35–40.

Langlois, D. E. 1986. "How You Can Transform That Staff Turkey into an Eagle," *The Executive Educator,* 8(3):22–23 + .

Langlois, D. E. 1986. "The Sky Won't Fall If Teachers Evaluate Principal Performance," *The Executive Educator,* 8(3):19–20.

Langlois, D. E. and M. R. Colarusso. 1988. "Teacher Evaluation: No Empty Ritual," *The Executive Educator,* 10(3):32–33. Reprinted in *The Education Digest* (November 1988):13–15.

Langlois, D. E. and M. R. Colarusso. 1988. "Massaging Egos: Overrating Performance, *Pennsylvania Educational Leadership,* 7(Spring):9–32.

Manning, R. C. 1986. "Evaluation Strategies Can be Improved with Peer Observers," *School Administrator,* 43(1):1.

Roper, S. S. and Hoffman, D. E. 1986. "Collegial Support for Professional

Development: The Stanford Collegial Evaluation Program," *OSSC Bulletin,* 29(9):36 pages. Publication Sales, Eugene, OR.

Ruck, C. L. 1986. "Creating a School Context for Collegial Supervision: The Principal's Role of Contractor," *OSSC Bulletin,* 30(3):39 pages. Publication Sales, Eugene, OR.

Snyder, K. J. 1986. "Competency Development for Principals," Paper presented for the American Educational Research Association, San Francisco, CA (ERIC Document ED279067).

Tye, K. A. 1986. "Better Teaching Through Instructional Supervision: Policy and Practice," California School Boards Association, Sacramento, CA 95816.

CHAPTER THREE

Elements of a Performance Appraisal Program for the Management Team

THE performance appraisal program described in this chapter is designed to address the evaluation priorities identified in the last chapter while at the same time drawing upon the positive elements of administrative appraisal that have evolved over the past generation. We will begin by a discussion of the concept that the job description should be used as the baseline document for the appraisal process.

The management-by-objectives approach will be included through the use of Individual Administrative Objectives. A procedure will also be recommended whereby members of each level of the administrative team receive evaluative input from their subordinates as a regular part of the process. Examples of forms to be used and procedures to be followed will also be provided. A self-evaluation tool will be proposed to be completed by each member of the administrative team. The manner in which evaluation can be tied to the compensation program will be further developed in a subsequent chapter.

The Job Description

In the authors' experience, job descriptions often suffer the same fate as policy manuals or the notes that many of us bring back from educational conferences. They are filed away safely for further study at a future time and are seldom, if ever, referred to again. In the extreme, we have found cases where administrators were not aware that there was a job description for their position. How can this type of situation be allowed to exist, and what can be done about it?

The new administrator reacts initially to several sets of expectations that influence his or her activities and functions. The supervisor, usually the superintendent, will have described the job requirements and special areas of concern during the interview process. The former incumbent in the job will have emphasized certain aspects of the job and may have neglected others. Teachers and other staff members will have their own perceptions of the job and will try to project these expectations onto the new administrator. Finally, the new administrator will have his or her own agenda of priorities and expectations for the position.

Including the job description as a critical element in the performance appraisal process can give this document new life and meaning. The simple act of critically reviewing the job description as part of the performance appraisal revision process can be a revealing and profitable exercise. Each time the authors have conducted this process with their administrators, individuals have discovered several items on the job description that are no longer valid and have found other areas of current importance that are not listed or are underemphasized.

Once administrators learn that their job descriptions will play a significant role in their evaluation and subsequent compensation, their interest in developing a document that accurately reflects reality increases dramatically. Each administrator is careful to include all major elements of his or her job on the job description and is quick to eliminate items that are no longer relevant or that are the responsibility of someone else.

Incorporating the job description into the performance appraisal process ensures that it will become a living document, no longer relegated to an obscure filing cabinet. With the job description as a guideline, your administrators will invest time and energy across the spectrum of their responsibilities. No longer will they overemphasize their own areas of interest or those functions that are forced to their attention by teachers or supervisors.

The job description itself should be developed and expressed in a manner that will enable it to be directly used as part of the evaluation process. The specific tasks should be grouped under major headings, each of which can be assigned a numerical rating indicating level of achievement as part of the summative evaluation process. The descriptors in the job description should be sufficiently explicit so that the degree to which they are being achieved can be readily observed or reasonably inferred. An example of such a research-based generic job description for a school principal follows.

TECHNOMIC SCHOOL DISTRICT
Generic Job Description: School Principal

A. Curriculum Development, Supervision, and Evaluation
   1. Knows district and school curriculum, ensures teaching of the written curriculum, helps staff use curriculum resources.
   2. Participates in and leads curriculum development activities commensurate with school and district goals.
   3. Provides opportunities and encouragement for staff to increase program expertise.
   4. Utilizes appropriate school and community personnel in evaluating curriculum.
   5. Identifies curricular and extracurricular needs by analyzing current programs and student achievement.
   6. Regularly uses the results of student testing to identify problems and implement program improvements.

B. Student Assessment and Monitoring
   1. Emphasizes student achievement as the primary outcome of schooling.
   2. Systematically assesses and monitors student progress using objective and verifiable information whenever possible.
   3. Provides meaningful information to parents and others regarding student progress.
   4. Works with staff to systematically identify and respond to at-risk students.
   5. Makes referrals to appropriate community agencies when needed.

C. Student and Staff Relations
   1. Models good human relations skills; effectively interacts with others.
   2. Facilitates the development of human relations skills among staff.
   3. Solicits information from school personnel and community in gauging school climate.
   4. Recognizes efforts of students and teachers.
   5. Promotes improvement of student and staff self-image.

    6. Fosters collegial relationships with and among teachers and staff.

    7. Communicates high expectations for both staff and students and provides appropriate motivation.

D. Establishing an Effective Workplace

    1. Develops and maintains positive staff morale.

    2. Implements a discipline code that is fair and promotes orderliness and student learning.

    3. Defines and articulates a school philosophy with vision.

    4. Sets and meets appropriate school-wide objectives on an annual basis.

    5. Protects instructional time by minimizing interruptions to the instructional process and teacher paperwork.

    6. Minimizes student absenteeism.

    7. Demonstrates a sensitivity to the dynamics of power; shares decision making.

    8. Coordinates student and teacher schedules to promote central objectives and minimize conflict.

    9. Maintains high visibility in the school.

    10. Promotes a climate that balances openness and control.

    11. Provides for adequate supervision and acceptable student behavior at all school sanctioned or sponsored activities.

    12. Orients new teachers to school programs and available resources.

E. Professional Development

    1. Identifies, plans, and implements staff development programs in accordance with assessed needs.

    2. Plans and implements individualized instructional improvement programs when necessary.

    3. Effectively utilizes the expertise of school personnel, including self, in staff development, and in-service programs.

    4. Helps teachers develop and implement objectives for themselves and students.

    5. Provides opportunities for teachers to share and demonstrate successful practices.

    6. Provides space, time, consultants, and other assistance for teachers to develop new or special instructional materials.

F. Staff Supervision and Personnel Evaluation

    1. Plans and implements a systematic personnel evaluation program that the staff understands.

    2. Continually monitors and revises the evaluation system utilizing information from appropriate personnel.

    3. Writes thorough, defensible, and insightful evaluation reports.

    4. Demonstrates objectivity in personnel evaluation.

5. Makes personnel assignments based on a knowledge of employee's ability, qualifications, past performance, and school needs.
6. Recognizes and responds to borderline performance and recommends removal of unsatisfactory personnel.

G. Communications

1. Listens and responds appropriately to staff, student, and community concerns.
2. Respects differences of opinion and fosters open communication among staff.
3. Develops communications that reflect and support management team decisions and school board policies.
4. Communicates effectively with students individually, in groups, and in school assemblies.
5. Speaks and writes effectively.
6. Keeps the superintendent and other appropriate central office administrators informed of school activities and problems.
7. Communicates and works with central office, supervisory personnel, and other principals to share ideas, problems, expertise, resources, and personnel.

H. Decision Making and Problem Solving

1. Considers research when making decisions.
2. Considers alternatives and consequences in the decision-making process.
3. Makes decisions in a timely fashion and maximizes decision effectiveness by follow-up actions.
4. Clearly communicates decisions and rationale to all affected.
5. Seeks information from appropriate sources and strives for consensus in the decision-making process.
6. Identifies problem areas and seeks solutions before crisis situations develop.
7. Effectively delegates decision making and problem solving to appropriate personnel.
8. Implements needed changes with appropriate support of staff, students, and the community.

I. Community Relations

1. Interacts with school district and parent groups to promote positive outcomes.
2. Keeps the community informed about school activities through newsletters, attendance at parent meetings, and the like.
3. Encourages parent visits and involvement in decision making.
4. Seeks appropriate community involvement in decision making.

    5. Provides appropriate programs for community audiences.
    6. Effectively utilizes community resources and volunteers to promote student learning.

J.   Personal Development

    1. Develops skills through participation in professional activities and organizations.
    2. Keeps abreast of current changes and developments within the profession.
    3. Views self as a role model for expected staff behavior.
    4. Perceives self as a change agent; works for self and organizational renewal.

K. Building Management

    1. Establishes and maintains rules and procedures for student and staff safety.
    2. Provides an aesthetically pleasing environment in the school.
    3. Monitors plant, office, and equipment maintenance.
    4. Provides for timely repair of school facilities and equipment.
    5. Effectively copes with crises and emergencies.

L. Record Keeping and Financial Management

    1. Maintains accurate personnel, student, and fiscal records.
    2. Prepares accurate budgets and effectively monitors expenditures.
    3. Prepares required district reports accurately and efficiently.
    4. Handles routine administrative matters effectively.
    5. Anticipates future building and equipment needs; plans appropriate activities.

Individual Administrative Objectives

The second element in the administrative performance appraisal program is the development of Individual Administrative Objectives. This part of the process ensures that each administrator will be striving to achieve objectives that represent major initiatives significantly beyond the range of day-to-day activities. A reliance on the job description alone can subtly promote the development of a bureaucratic management mentality. Challenging and far-sighted objectives, however, can help the administrator develop his or her leadership skills and can foster significant educational improvements for the organization and the students.

All of us can reflect on our careers as teachers and remember the principal or assistant principal who was a master of administrivia. He or she would observe one of your lessons and comment that the window blinds were crooked. Or you might have received a memo complaining about your excessive use of copying paper. In the faculty room your colleagues often referred to this administrator as "the master of micromanagement."

Administrators of this type are the victims of administrative gridlock. They have become so consumed by mundane details that they lack any sense of a larger vision, an absolute prerequisite for effective leadership. They not only do not see the big picture, they don't know that there is one. Administrators who are permitted to persist at this level of mediocrity are ill-served by their superintendent and school board.

A vigorous program of Individual Administrative Objectives can rejuvenate such lackluster administrators. IAOs are the principal vehicles by which a superintendent can develop the leadership abilities and skills of his or her subordinates. This management-by-objectives approach represents a critical element in any effective performance appraisal program. A seasoned administrator in particular, who has essentially mastered the elements of his or her job description, can continue to grow professionally and excel as a leader through involvement in an IAO process.

In developing Individual Administrative Objectives, the administrator should strive to state the objectives in terms that are clear and specific. The degree to which an objective has been achieved should be measurable, and the data to be utilized should be referenced in the objective. Specific steps that the administrator will follow to implement the objective should also be included in the commentary on the objective. Three or four objectives of the kind described above are the maximum that should be attempted in any one year.

The positive impact that a vibrant administrative objectives program can have on a school district can be dramatic. If each administrator in a district with ten administrators were to achieve three significant objectives in a year, the district total for these leadership initiatives would be thirty. The cumulative positive effect of such a program over several years could literally transform a school district.

More important than the specific objectives themselves, however, is the message conveyed by the process — that the district values and rewards goal-directed behaviors by its administrators. The constant identification of new objectives also mitigates against complacency in an organization and establishes a standard of continual improvement as an administrative norm.

Three examples of Individual Administrative Objectives that generally conform to the criteria outlined above are discussed in the following paragraphs. All of these objectives are for a high school principal. The first one is process oriented in that it outlines a series of activities that will be undertaken to work toward achievement of the objective.

OBJECTIVE— The principal will, with the help of department chairpersons and the assistant superintendent for curriculum and instruction, critically assess the current status of the written curriculum in each subject area and prepare a report for the superintendent summarizing his findings and recommendations.

The major activities relating to this objective are as follows:

(1) Each department chairperson will submit a written and oral outline to the other department chairpersons and the principal. The assistant superintendent for curriculum and instruction will be invited to attend these meetings and will be given copies of all reports. These reports will include, but will not be limited to, the following:

  • overview of subjects taught, grade level taught, course sequence, etc.
  • description of changes in course offerings over the past three years
  • recommendations and suggestions including appropriate timelines

(2) Each department chairperson will develop one or two short-range objectives (one to three years) and one or two long-range goals (four to six years) based on the needs identified in the above report. Such goals and objectives should be as specific as possible and should include possible budget and program implications.

(*3*) The principal will submit a report to the superintendent by May of the current school year summarizing the findings and recommendations that derive from this curriculum review process.

The second objective detailed below is basically similar to the first, with the additional component of gathering statistical and other hard data as part of the process.

OBJECTIVE— *The principal will study the nature and characteristics of a sample of dropouts from Technomic High School and will propose program changes to better serve such students based on an analysis of the data collected.*

The specific activities to be undertaken in support of this objective are as follows:

(*1*) Analyze the nature and characteristics of a sample of dropouts for school years 1986-1991 by examining the following information:
- family history and current family dynamics
- student participation in school activities
- attendance, suspension, and detention data
- academic performance
- other appropriate data

(*2*) Review the research on effective dropout prevention programs and visit nearby schools that have programs in place.

(*3*) Assess three years of data on the current dropout prevention program at Technomic, including analysis of the program's impact on the following areas:
- attendance rates
- grade-point averages
- suspension and detention rates
- dropout rates
- other appropriate data

(*4*) Based on the above research and study, develop a systematic plan to increase successful interventions with students who are in danger of dropping out of school.

The above two objectives specify in detail the steps that will be followed to achieve the objective. It is possible to write a more generalized objective that contains the basic elements of a good Individual Administrative Objective. The objective below is an example of this more generalized approach.

OBJECTIVE—The high school grading and report system will be studied and recommendations will be made to modify the existing system to better meet the needs of the students and the high school.

A committee of volunteer staff members will review existing data regarding grade distributions by subject for all students and grade levels. A survey will be completed by all staff members, identifying their criteria and procedures for assigning grades. Additional input on the topic will be solicited from parents and students. Samples of grading systems used by nearby school districts will be reviewed by the committee. A summary report and recommendations for change will be prepared by the principal in consultation with the committee and will be presented to the faculty for consideration by May of the current school year.

Subordinates Evaluate Their Supervisors

The third major element in the appraisal program should be some formal provision for the input of subordinates in evaluating the performance of their supervisors. This evaluation should be limited to those aspects of the supervisor's responsibilities that directly affect the subordinate and for which the subordinate has direct evidence or experiences.

There is an old military adage that the commander who gets too far ahead of his troops is likely to get shot in the rear. This saying applies also to the relationship between a school administrator and his or her subordinates. Due to the very public nature of the schools, teachers and other staff members are uniquely able to undercut the effectiveness and reputation of their administrator both in the school and in the community.

Introducing a formal process to measure the opinions and judgments of subordinates about the quality of administrative leadership can serve as a valuable communication link between the administrator and his or her subordinates. In our experience, however, it is the less effective administrators who are the most resistant to soliciting such information from their subordinates.

Dealing effectively with this issue requires both tact and diplomacy on the part of the superintendent. As a first step, the superintendent should provide an example to his or her immediate subordinates by developing a formal process for soliciting their opinions regarding the performance of the superintendent. Secondly, the superintendent should require that each administrator solicit the opinions of his or her staff members regarding administrative performance. This should be accompanied by repeated assurances that the results of these opinion surveys are not expected to be shared with the superintendent.

This evaluation instrument should be stated in terms of observable behavior and should be rather brief and easy to administer and interpret. The data should be gathered and tallied by someone other than the supervisor. The administrators should be required to collect these data as part of their evaluation process and should be encouraged, but not required, to share this information with their supervisors.

An example of such a generic instrument for use in the evaluation of a school principal follows.

TECHNOMIC SCHOOL DISTRICT
Teachers Evaluate Their Principal

Please circle the number that best represents your opinion of the principal's performance. Circle 5 if the statement is always true, circle 4 if it's almost always true, circle 3 if it's frequently true, circle 2 if it's occasionally true, circle 1 if it's never true, and circle NA if it's not applicable.

1. Knows district and school curriculum, ensures teaching of the written curriculum, helps staff use curriculum resources.   1 2 3 4 5 NA

2. Provides opportunities and encouragement for staff to increase program expertise.   1 2 3 4 5 NA

3. Utilizes appropriate school and community personnel in evaluating curriculum.   1 2 3 4 5 NA

4. Emphasizes student achievement as the primary outcome of schooling.   1 2 3 4 5 NA

5. Works with staff to systematically identify and respond to at-risk students.   1 2 3 4 5 NA

6. Models good human relations skills; effectively interacts with others.   1 2 3 4 5 NA

7. Facilitates the development of human relations skills among staff.   1 2 3 4 5 NA

8. Solicits information from school personnel and community in gauging school climate.   1 2 3 4 5 NA

9. Recognizes the efforts of students and teachers.   1 2 3 4 5 NA

10. Communicates high expectations for both staff and students and provides appropriate motivation.   1 2 3 4 5 NA

11. Develops and maintains positive staff morale.   1 2 3 4 5 NA

12. Defines and articulates a school philosophy with vision.   1 2 3 4 5 NA

13. Protects instructional time by minimizing interruptions to the instructional process and teacher paperwork.   1 2 3 4 5 NA

14. Demonstrates a sensitivity to the dynamics of power; shares decision making.   1 2 3 4 5 NA

15. Coordinates student and teacher schedules to promote central objectives and minimize conflict.   1 2 3 4 5 NA

16. Maintains high visibility in the school.   1 2 3 4 5 NA

17. Orients new teachers to school programs and available resources.   1 2 3 4 5 NA

18. Effectively utilizes the expertise of school personnel, including self, in staff development, and in-service programs.   1 2 3 4 5 NA

19. Helps teachers develop and implement objectives for themselves and students.   1 2 3 4 5 NA

20. Provides opportunities for teachers to share and demonstrate successful practices.   1 2 3 4 5 NA

21. Plans and implements a systematic personnel evaluation program that the staff understands.   1 2 3 4 5 NA

22. Demonstrates objectivity in personnel evaluation.   1 2 3 4 5 NA

23. Respects differences of opinion and fosters open communication among staff.   1 2 3 4 5 NA

24. Makes decisions in a timely fashion and maximizes decision effectiveness by follow-up actions.   1 2 3 4 5 NA

25. Identifies problem areas and seeks solutions before crisis situations develop.   1 2 3 4 5 NA

Administrator Self-Evaluation

The final element in the evaluation process should be a self-evaluation by the administrator. Most people are able to acknowledge and accept areas for professional growth that they themselves have identified. Through this self-evaluation mechanism the supervisor will find that the subordinate will often introduce topics for discussion which closely match areas of concern to the supervisor.

This dynamic of having the subordinate, in effect, ask for assistance in certain areas places the supervisor more in the role of a coach than of a traditional boss critiquing the work of a subordinate. The person being evaluated can either be asked to assess himself using the job description form that is to be used by the supervisor or can be asked to complete a more open-ended and generic instrument for administrative evaluation. A somewhat less formal procedure is to request that each administrator arrive at the appraisal conference prepared to discuss several items taken from the job description that are of current interest or concern to the administrator. The self-evaluation element will, of course, be an integral part of the discussion between the administrator and supervisor as they jointly review progress toward the achievement of the Individual Administrative Objectives.

An example of the open-ended type of form follows.

TECHNOMIC SCHOOL DISTRICT
Administrator Self-Evaluation

Name_____**Position**_____**Date**_____

Directions: **Please read and react to the following items as appropriate for your position prior to your conference with the Superintendent. These questions will be used as part of the discussion.**

1. Professional Growth—What have you done over the year?

_____

_____

2. In what school-related programs and activities have you participated during the past year, and how did they benefit you and the district?

_____

3. During the past year, what working relationship did you develop with the following? Why?
   A. Non-Professional Staff:

   _____

   B. Teacher:

   _____

   C. Administrators/Supervisors:

   _____

   D. Students:

   _____

   E. Parents:

   _____

   F. School Board:

   _____

4. In what way during the past year have you improved communication with the following?
   A. Non-Professional Staff:

   _____

   B. Teachers:

   _____

   C. Students:

   _____

   D. Other Administrators/Supervisors:

   _____

   E. Parents:

   _____

   F. School Board:

   _____

5. Area of Instruction:
   A. How many classroom observations did you complete this year?

   How did your visits help the teachers and students?

   _____

What did you gain?

_____

B. What were the program innovations studied during the year?

_____

C. What did you do this year to improve your staff meetings and in-service programs?

_____

_____

D. What contributions did you make this year to the K-12 concept?

_____

_____

E. What was your greatest instructional achievement this year?

_____

_____

6. Area of Management:
   A. Did you carry on your special assignments without having to be prodded?

   _____

   _____

   B. What "difficult decisions" did you face during the school year?

   _____

   _____

   C. What role did you play in the following areas?
      • Board policy development and implementation:_____
      • Budget input and budget administration:_____

7. If you had complete authority, what changes would you make in the following areas?
   A. Physical Plant

   _____

   B. Custodial Service/Maintenance

   _____

   C. Supplies/Equipment

   _____

   D. Secretarial Service

   _____

E. Lunch Program

_____

F. Specialists

_____

G. Administrative Decision-Making Authority and Responsibility

_____

H. Athletic Program

_____

I. Music Program

_____

J. Staff

_____

K. Other

_____

_____

8. What has been your greatest contribution to the school district during the past year?

_____

_____

9. What do you consider to be the area of your performance during the past year that is in need of improvement?

_____

_____

10. What do you intend to do to improve your performance next year?

_____

_____

_____

11. What special thing do you feel you can contribute to the Management Team next year?

_____

_____

12. Could anyone have been of more assistance to you during this past year?
    When?_____
    Who?_____
    How?_____

13. What are your hopes for improving your building/educational program next year?

_____

_____

Informal Components of Performance Appraisal

The concept of coaching as part of the appraisal process can be further enhanced through the inclusion of two additional activities that can be considered to be informally related to performance appraisal. The first of these ideas is borrowed from the concept of mentoring as currently employed in teacher induction programs in many states and school districts. The second suggestion is to provide some structured opportunities for a new administrator to directly access the accumulated experience and insights of his or her superintendent. These two suggestions are applicable to administrators during their first year in a school district.

Each administrator new to the district should be paired with a volunteer experienced administrator. This experienced administrator would not be in a direct supervisory relationship with the new administrator. The initial pairing would be determined by the superintendent, who would be in the best position to match the perceived needs of the new administrator with the talents and skills of the seasoned administrator. The new administrator and the mentor would then meet periodically to discuss general topics relating to school administration, with an emphasis on succeeding as an administrator in their particular school district.

The second suggested informal activity involves the provision for structured opportunities for the new administrator to meet with the superintendent outside of the formal evaluation conference schedule. The purpose of these meetings would be to provide the new administrator with district-wide or even society-wide perspectives on the problems and challenges of school administration. Such opportunities could help to develop administrators who can see the larger picture and who will be better able to function in their particular role in the organization.

One of the authors made it a practice to meet informally with new administrators during their first year in the school district. These far-ranging discussions not only provided the new person with the larger perspective of the superintendency, but also provided the superintendent with first impressions and judgments on school district operations from someone who was experiencing the culture and procedures of school district operation for the first time.

Opportunities to meet informally with the superintendent are particularly appreciated by new assistant principals. These new administrators, typically mired in the trench warfare of student discipline, savor the chance to discuss the larger issues of school administration with the superintendent.

Both of these ideas, although not intended as formal parts of the appraisal process, can nonetheless directly influence the growth and development of the administrator. In this sense they represent worthwhile supplements to a well-designed performance appraisal program.

Each of the performance appraisal elements described in this chapter will be applied to specific administrative positions that will be discussed in detail in the chapters that follow. The consistent attention to these elements throughout all levels of the organization will ensure unity and cohesiveness for the entire performance appraisal program within a school district.

Suggested Readings

Bailey, G. D. 1984. "Faculty Feedback for Administrators: A Means to Improve Leadership Behavior," *NASSP Bulletin*, 68(1):5–9.

Cawelti, G. 1984. "Behavior Patterns of Effective Principals," *Educational Leadership*, 41(2):3.

Deal, T., S. Dornbusch, and R. Crawford. 1977. "Villans as Victims: Evaluating Principals," *Phi Delta Kappan* (December): 273.

Duke, D. L. and R. Stiggins. 1985. "Evaluating the Performance of Pincipals: A Descriptive Study," *Educational Administration Quarterly*, 21(Fall):71–98.

Ernest, B. 1985. "Can You Eat? Can You Sleep? Can You Laugh? The Why and How of Evaluating Principals," *The Clearing House*, 85(3):290–292.

Ginsberg, R. and P. Berry. 1990. "The Folklore of Principal Evaluation," *Journal of Personnel Evaluation in Education*, 3(3):205–230.

Grabinski, D., E. Look, and J. Sweeney. 1985. "Follow These Four Steps to Solid Administrator Evaluation," *The Executive Educator*, 7(April): 25–30.

Hallinger, P and J. F. Murphy. 1987. "Assessing and Developing Principal Instructional Leadership," *Educational Leadership,* 45(9):54–61.

Hallinger, P. and J. F. Murphy. 1985. "Assessing the Instructional Management Behavior of Principals," *The Elementary School Journal,* 86(11): 217-247.

Harrison, W. C. and K. D. Peterson. 1988. "Evaluation of Principals: The Process Can Be Improved," *NASSP Bulletin,* 72(5):104.

Herman, J. J. 1988. "Evaluating Administrators—Assessing the Competencies," *NASSP Bulletin,* 72(5):5-6, 8-10.

Klopf, G. J., E. Scheldon, and K. Brennan. 1982. "The Essentials of Effectiveness: A Job Description for Principals," *Principal,* 61(3):35-39.

Langlois, D. E. 1986. "The Sky Won't Fall If Teachers Evaluate Principal Perfomance," *The Executive Educator* (3):19-20.

Lindahl, R. A. 1987. "Evaluating Principal's Performance: An Essential Step in Promoting School Excellence," *Education,* 108(Winter):204-211.

Look, E. and R. Mannatt. 1984. "Evaluating Principal Performance with Improved Criteria," *NASSP Bulletin,* 68(12):76-81.

Manasse, A. L. 1985. "Improving Conditions for Principal Effectiveness: Policy Implications of Research," *The Elementary School Journal,* 85(1): 439-463.

Murphy, J., P. Hallinger and K. D. Peterson. 1985. "Supervising and Evaluating Principals: Lessons from the Effective Districts," *Educational Leadership,* 43(10):78-82.

Redfern, G. B. 1986. "Techniques of Evaluation of Principals and Assistant Principals: Four Case Studies," *NASSP Bulletin,* 70(2):66-74.

Solomon, G. 1983. "Teachers Rating Principals: A Sobering Experience," *Principal,* 62(3):14-17.

Stow, S. B. and R. P. Manatt. 1982. "Administrator Evaluation Tailored to Your District or Independent School," *Educational Leadership,* 39(2): 353-356.

Vann, A. S. 1989. "When Teachers Grade the Principal," *Principal,* 68(3): 46-47.

CHAPTER FOUR

Performance Appraisal of Building Administrators

EACH level in the organizational structure provides special challenges and conditions for conducting the performance appraisal process. We will deal specifically in this chapter with the evaluation of administrative performance at the building level. In this discussion we will be speaking primarily of building principals, although many of the points discussed would also hold true for assistant principals. We will begin with a discussion of special factors in the worklife of a building principal.

Special Aspects of the Building Principalship

The building principal, compared with other administrators, has his or her own constituency and has high visibility in the community. He or she must interact on a more constant basis with parents and students who are not part of the educational fraternity. This situation provides the principal with an independent power base and with the potential to powerfully influence public opinion about his or her school and the school district.

It is natural for principals to interpret school district directives and policies in terms of how they will affect their particular school. Principals frequently see themselves as advocates for their schools in determining how scarce resources should be allocated among the schools in the district. Conversely, it is less common for a principal to have a well-developed district-wide perspective on most school issues. A building principal willing to adopt a district-wide perspective on critical issues should be prized by a superintendent and school board.

Most frequently, parents and students think of the principal when they hear or read the term *school administrator*. Their interactions with the principal regarding academic, extracurricular, or discipline concerns greatly influence their perception of the competence of the management of their school district. The high school principal usually has the highest public visibility and interaction with the community of any school district administrator.

The principal is also the major point of contact between the teachers and the school district administration. The skill of the various school principals in the supervision of their staffs significantly affects the morale and general attitude of teachers toward the management of the school district. Good principals can partially mitigate the negative effects of a poor central administration. Poor principals, on the other hand, can create a barrier to positive staff relations that is virtually impossible to overcome from a district level.

This is a good place to discuss the loyalty of a principal to his or her administrative colleagues and to the superintendent. Both authors have worked with otherwise effective principals who have tarnished their own reputations and compromised their own effectiveness through unprofessional and unwarranted criticism of their colleagues. This type of behavior is insidious and yet is often difficult for a superintendent to detect.

The principal who constantly criticizes administrative peers is suffering from a form of sibling rivalry that derives from his or her own insecurities and inadequacies. The superintendent can only become aware of such unfair criticisms or misinformation through informal conversations with teachers or other staff members. Unless a superintendent has a ready rapport with staff members at all levels of his organization, he or she might remain unaware that this problem exists.

The same principals who frequently criticize other administrators in the district are also the most likely to be frequent critics of central office staff and the superintendent. While it is certainly desirable for a principal to protect the interests of his or her school and staff vis-a-vis central office, this should not bring about an "Us versus Them" mentality at the building level. A weak principal often finds it expedient to distract the dissatisfactions of his or her staff toward a perceived common enemy at the district office level.

The subject of professional loyalty among all administrators in the school district should be addressed either directly or indirectly as part of the principal evaluation process. The superintendent must emphasize to the principal that, although he or she is the leader of an individual school, the school is part of a larger system that relies upon cooperation among administrators to meet the educational goals of the district.

The goals and objectives that a principal chooses to pursue must, of necessity, be achieved through and with other people. To improve the attendance rate, for example, the principal will need to gain the active support of the office staff, teachers, students, and their parents. Attendance is also related to the quality of the instructional program and the extent to which individual teachers can succeed in helping reluctant students to become truly engaged in the learning process. Thus the achievement of a seemingly straightforward objective such as improving student attendance is actually a complex undertaking requiring much coordination, analysis, and staff cooperation.

Supervising the Building Administrator

The above paragraphs illustrate that the success of a building administrator depends in large measure on his people skills, rather than primarily on management skills or knowledge of curriculum and instruction. Thus any appraisal process for a building administrator should include extensive personal observation by the supervisor of the principal and his interactions with staff members at the building level.

The direct supervisor of the building principal, which should be the superintendent in all but the largest districts, should make frequent visits to the schools. By observing the day-to-day activities within the school, the supervisor will gain valuable insights into the leadership style and effectiveness of the principal. As superintendents, both authors included all-day shadowing of their principals as part of their supervisory repertoire. The climate of a school is powerfully influenced by the leadership of the principal. The supervisor can generally infer that the positive and negative aspects of school climate that he or she observes have been affected by the principal.

The supervisor should be looking for a school with a relaxed yet businesslike atmosphere. Students and teachers should demonstrate a high level of engagement in the learning process. In a healthy school environment, teachers will be willing to extend themselves beyond their classroom duties and participate fully in advancing the mission and goals of the school.

Teachers should be willing to volunteer for committee work and for leadership roles in staff development and extracurricular activities. Staff members should exhibit a willingness to explore new ideas and programs to improve the performance of the school. There should be evidence of continuous staff involvement in school improvement and program evaluation activities.

In their own evaluations of building principals, the authors sought to emphasize the principal's role as an instructional leader rather than as a building manager. Such an emphasis requires that the superintendent assign priority status to certain items on the job description and specific IAOs relating directly to the instructional program.

In evaluating the principal as instructional leader, the authors typically considered the following areas:

(1) *Selection and Evaluation of Teachers:* If the quality of teaching is the most important element in an effective school district, then surely the principal is critically placed to attract and develop teaching talent. While we rightly applaud the ability of a principal to work effectively with the few marginal teachers, we should be even more appreciative of his or her ability to convert the more numerous mediocre teachers into star performers. Teacher selection, development, and evaluation are key skills of the principal as instructional leader.

(2) *Curriculum and Program Changes:* Even school districts with central office personnel for Curriculum and Instruction must still rely on the skill of the principal to bring about curriculum change and program improvement. The principal interacts with the teachers daily and is in a position to allocate instructional resources in support of new curriculum initiatives. No significant program improvements will ultimately succeed without the active support of the building principal.

(3) *Allocation of Instructional Resources:* This function establishes the priorities of the school with respect to allocation of monies for textbooks, other instructional materials, and instructional

equipment. These decisions materially affect the breadth, depth, and quality of the educational program.

(4) *Construction of a Master Schedule for Students and Teachers:* Through this mechanism the principal determines how hundreds of people will each be spending about a thousand hours of their time in the course of a school year. This often underrated administrative function can have a profound effect on the educational achievement of students.

The soft data on the principal's performance gathered through the school visitation process will serve as a valuable complement to the formal data provided by the administrator as part of the appraisal process. In the case of principal evaluations, the supervisor's experiences during his or her visits to the buildings will provide the necessary background for properly interpreting the formal reports offered by the principal in responding to his or her job description and individual administrative objectives.

In evaluating principals we should remember that crisis management is a constant of the public school principalship. The superintendent must be careful to avoid placing too much significance on the manner in which a principal responds to a particular crisis situation.

It sometimes happens, however, that a given issue can dominate an entire year of a principal's time and energy. A case in point occurred in the district of one of the authors when a small group of parents objected to a classroom drug and alcohol program being used in the middle school. The parents, as well as non-parent community members, objected to the use of specific commercially available materials because of their assertion that they had been misused in other school districts. They further claimed that the school was teaching students to reject their parents' values and to substitute for them a moral relativism that they identified as secular humanism. This group of critics also hinted darkly that a New Age philosophy was being taught to the students.

The school district initially responded to the challenge through meetings and correspondence between the curriculum director and the objecting citizens. The controversy soon spread directly to the school, however, requiring the involvement of the principal. Over a period of several months, this principal responded to questions and concerns from the original objectors as well as those of other parents who became worried about the allegations. Thus, this principal found that large portions of his time and energy needed to be directed toward this issue. This was true even though the controversy was created by only a handful of people, none of whom had children participating in the program.

Despite the best efforts of the curriculum director and the principal and staff, the problem did not resolve itself. Throughout the school year this issue kept resurfacing and demanding even greater time and energy resources on the part of the principal. As his supervisor, the superintendent had to evaluate the manner in which the principal dealt with this critical issue, make allowances for somewhat diminished attention to other areas of responsibility, and yet arrive at an overall evaluation that adequately considered all aspects of the principal's responsibilities.

In our experience, the principalship is the level where there seems to be the greatest need to factor the special circumstances of a particular school year into the evaluation equation. Having said this, however, we also offer the caution that the principal should not be permitted to adopt a fire brigade mentality whereby the constant problems of the moment effectively prevent long-term planning and the achievement of major objectives.

On the following pages, examples of job descriptions have been provided for the high school principal and the assistant high school principal in the fictitious Technomic School District. Both of these documents provide extensive lists of possible descriptors for these positions. An actual job description would contain many, but not all of the descriptors listed. Also included is an elementary principal's job description (which is also used as a self-evaluation form), job descriptors, and a rating scale from the Cornwall-Lebanon School District in Lebanon, Pennsylvania. Several examples of sets of Individual Administrative Objectives suitable for building administrators have also been provided.

TECHNOMIC SCHOOL DISTRICT
Job Description: High School Principal
Reporting Relationship
    1. Reports to superintendent of schools.

Primary Functions
1. The high school principal is responsible for the supervision of all teaching, administrative, clerical, and service staff members in his or her school.
2. The high school principal is responsible for the supervision of the education program in his or her school, particularly in the areas of curriculum and instruction.
3. The high school principal maintains a close relationship with the superintendent to assure that the programs and policies of the school board are carried out in an efficient and uniform manner.

Performance Responsibilities

Curriculum Development, Supervision, and Evaluation
1. Knows district and school curriculum, ensures teaching of the written curriculum, helps staff use curriculum resources.
2. Participates in and leads curriculum development activities commensurate with school and district goals.
3. Provides opportunities and encouragement for staff to increase program expertise.
4. Utilizes appropriate school and community personnel in evaluating curriculum.
5. Identifies curricular and extracurricular needs by analyzing current programs and student achievement.
6. Regularly uses the results of student testing to identify problems and implement program improvements.
7. Effectively supervises curriculum coordinators.

Student Assessment and Monitoring
1. Emphasizes student achievement as the primary outcome of schooling.
2. Systematically assesses and monitors student progress using objective and verifiable information whenever possible.
3. Provides meaningful information to parents and others regarding student progress.
4. Works with staff to systematically identify and respond to at-risk students; makes referrals to appropriate community agencies when needed.

Student and Staff Relations
1. Models and facilitates good human relations skills; effectively interacts with others.
2. Solicits information from school personnel and community in gauging school climate.
3. Recognizes efforts of students and teachers.
4. Promotes improvement of student and staff self-image.
5. Fosters collegial relationships with and among teachers and staff.

    6. Communicates high expectations for both staff and students and provides appropriate motivation.

    7. Understands and manages contract between the Technomic Teachers Association and the Technomic School District.

Establishing an Effective Workplace

    1. Develops and maintains positive staff morale.

    2. Implements a discipline code that is fair and promotes orderliness and student learning.

    3. Defines and articulates a school philosophy with vision.

    4. Protects instructional time by minimizing interruptions to the instructional process, student absenteeism, and teacher paperwork.

    5. Coordinates student and teacher schedules to promote central objectives and minimize conflict.

    6. Maintains high visibility in the school.

    7. Promotes a climate that balances openness and control.

    8. Provides for adequate supervision and acceptable student behavior at all school sanctioned or sponsored activities.

    9. Orients new teachers to school programs and available resources.

Staff Supervision and Personnel Evaluation

    1. Plans and implements a systematic personnel evaluation program that staff understand.

    2. Continually monitors and revises the evaluation system utilizing information from appropriate personnel.

    3. Writes thorough, defensible, and insightful evaluation reports.

    4. Demonstrates objectivity in personnel evaluation.

    5. Makes personnel assignments based on a knowledge of employee's ability, qualifications, past performance, and school needs.

    6. Recognizes and responds to borderline performance and recommends removal of unsatisfactory personnel.

Communications and Community Relations

    1. Listens and responds appropriately to staff, student, and community concerns.

    2. Respects differences of opinion and fosters open communication among staff.

    3. Develops communications that reflect and support management team decisions and school board policies.

    4. Communicates effectively with students individually, in groups, and in school assemblies.

    5. Speaks and writes effectively.

    6. Keeps the superintendent and other appropriate central office administrators informed of school activities and problems.

    7. Communicates and works with central office, supervisory personnel, and other principals to share ideas, problems, expertise, and personnel.

    8. Interacts with school district and parent groups to promote positive outcomes.

9. Keeps the community informed about school activities through newsletters, attendance at parent meetings, etc.
10. Encourages parent visits and involvement in decision making.
11. Seeks appropriate community involvement in decision making.
12. Provides appropriate programs for community audiences.
13. Effectively utilizes community resources and volunteers to promote student learning.

Decision Making and Problem Solving
1. Considers research when making decisions.
2. Considers alternatives and consequences in the decision-making process.
3. Makes decisions in a timely fashion and maximizes decision effectiveness by follow-up actions.
4. Clearly communicates decisions and rationale to all affected.
5. Seeks information from appropriate sources and strives for consensus in the decision-making process.
6. Identifies problem areas and seeks solutions before crisis situations develop.
7. Effectively delegates decision making and problem solving to appropriate personnel.
8. Implements needed change with appropriate support of staff, students, and community.

Professional Development
1. In cooperation with staff development committee, identifies, plans, and implements staff development programs in accordance with assessed needs.
2. Plans and implements individualized instructional improvement programs when necessary.
3. Effectively utilizes the expertise of school personnel, including self, in staff development, and in-service programs.
4. Helps teachers develop and implement objectives for themselves and students.
5. Provides opportunities for teachers to share and demonstrate successful practices.
6. Provides space, time, consultants, and other assistance for teachers to develop new or special instructional materials.
7. Develops skills through participation in professional activities and organizations.
8. Keeps abreast of current changes and developments within the profession.
9. Views self as a role model for expected staff behavior.
10. Perceives self as a change agent; works for self- and organizational renewal.

Building Management, Record Keeping, and Financial Management
1. Establishes and maintains rules and procedures for student and staff safety.

2. Provides an aesthetically pleasing environment in the school.
3. Monitors plant, office, and equipment maintenance.
4. Provides for timely repair of school facilities and equipment.
5. Effectively copes with crises and emergencies.
6. Maintains accurate personnel, student, and fiscal records.
7. Prepares accurate budgets and effectively monitors expenditures.
8. Prepares required district reports accurately and efficiently.
9. Handles routine administrative matters effectively.
10. Anticipates future building and equipment needs; plans appropriate activities.

TECHNOMIC SCHOOL DISTRICT
Job Description: Assistant High School Principal

Reporting Relationship
1. Reports to the high school principal.

Primary Functions
1. The assistant high school principal provides the necessary background and leadership for organizing and administering the high school.
2. The assistant high school principal assists the principal in the planning, staffing, budgeting, and evaluation of the program.
3. The assistant high school principal has the primary administrative responsibility for the ninth and tenth grade program.

Performance Responsibilities

Curriculum Development, Supervision, and Evaluation
1. Knows district and school curriculum and assists the principal in monitoring the teaching of the written curriculum.
2. Participates in curriculum development activities commensurate with school and district goals.
3. Provides opportunities and encouragement for staff to increase program expertise.
4. Collaborates with department heads and teachers in order to assist the principal in the development, revision, and evaluation of the curriculum.

Student Assessment and Monitoring
1. Emphasizes student academic achievement and social and emotional development as the primary outcomes of the high school program.
2. Provides for the appropriate academic placement for the ninth and tenth grade students.
3. Participates in Child Study Team meetings to identify at-risk students and to develop appropriate programmatic plans to meet their needs.
4. Administers the distribution of progress reports and report cards.

5. Participates in the Student Assistance Program and serves as liaison with district administration regarding SAP.
6. Organizes and coordinates mental health and drug and alcohol prevention programs.
7. Participates in Multi-Disciplinary Team meetings when deemed appropriate.
8. Administers the Gifted Program.

Student and Staff Relations

1. Demonstrates positive and supportive interactive skills and facilitates the development of good collegial relations skills among staff by providing opportunities for professional collaboration.
2. Recognizes the efforts of students and teachers through written and verbal communications.
3. Coordinates and monitors the Student Mentor Program.
4. Serves as administrative liaison with the Student Council. Conducts monthly meetings with Student Council officers.
5. Encourages and supports teacher- and student-initiated projects that foster a positive school climate.
6. Identifies and coordinates appropriate assembly programs.

Establishing an Effective Workplace

1. Collaborates with the principal to develop and maintain positive staff morale.
2. Implements the district's Discipline Code of Conduct and works with the principal to make code revision recommendations to the District Discipline Committee when deemed appropriate.
3. Coordinates and monitors the In-school Suspension Program and the Administrative Detention Program.
4. Plans, implements, and evaluates absence and lateness policies and procedures with the principal to effect improvement in student attendance.
5. Assists the principal in setting and meeting appropriate school-wide objectives on an annual basis.
6. Assists the principal in organizing and supervising school-related activities, keeping interruptions to the instructional process at a minimum.
7. Maintains high visibility and accessibility in the school.
8. Assists the principal in developing the master schedule.
9. Administers and monitors teacher supervisory assignments and duties.
10. Works with the athletic director to coordinate student eligibility and other student athletic concerns and problems.
11. Shares the responsibility of discipline and student control with the principal.

Professional Development

1. Assists the principal in identifying, planning, and implementing staff development programs in accordance with assessed needs.

2. Collaborates with the principal to develop and implement individual-ized instructional improvement programs when necessary.
3. Works with teachers to develop and implement objectives for themselves and students through conferencing and information sharing.
4. Provides opportunities for teachers to share and demonstrate successful practices.

Staff Supervision and Personnel Evaluation
1. Conducts classroom observations and conferences and writes thorough and meaningful reports.
2. Assists the principal in making personnel assignments based on knowledge of employee's ability, qualifications, past performance, and school needs.
3. Keeps the principal informed of observed borderline performance and recommendations made.
4. Collaborates with the principal regarding the evaluation system and assists the principal in its implementation.

Communications
1. Develops communications that reflect and support management team decisions and school board policies:
   • annual review and revision of the Teacher Handbook
   • annual review and revision of the grade 9-12 informational folder
   • periodic correspondence to staff and/or students
2. With the principal's knowledge, keeps district office administrators informed of district-related activities and problematic situations.
3. Conferences with students individually and in groups as needed and keeps students informed of planned activities and procedural issues.
4. Communicates and works with district office, supervisory personnel, and other administrators to share ideas, problems, expertise, resources, and personnel.
5. Provides ongoing information to the principal regarding school-related activities, programs, and problematic situations.

Decision Making and Problem Solving
1. Assists the principal in the development of school policies and procedures giving due consideration to related research, alternatives available, and possible consequences.
2. Demonstrates sound judgment in decision making in areas of responsibility and establishes and communicates procedural requirements to implement action plan.
3. Consults the principal and other appropriate sources prior to making important school-related decisions.
4. Assists the principal in the interviewing of candidates for professional positions in the schools.

Community Relations
1. Serves as liaison between the high school and the local police and fire departments to promote communication and cooperation.

2. Assists in developing a sound school-community relationship by:
   - providing information
   - working with student organizations to foster community involvement
   - attending meetings
   - responding to expressed community needs
3. Works with the principal and guidance personnel to organize parent visitations and to consider parent concerns and recommendations when making decisions.
4. Assists the principal in providing appropriate programs for parents and community audiences.

Personnel Development

1. Participates in professional organizations, workshops, and conferences to gain information regarding the latest developments in the various aspects of administration and instructional programming.
2. Represents the principal, upon request, at meetings and functions that the principal is unable to attend.
3. Recognizes own areas needing improvement and works towards self-growth and -improvement.

Building Management

1. Assumes leadership of the school in the absence of the principal.
2. Collaborates with the principal to establish and maintain rules and procedures for student and staff safety.
3. Assists the principal in monitoring plant, office, and equipment maintenance and submits appropriate work orders to auxiliary services in a timely fashion.
4. Works with appropriate support personnel to organize and maintain locker assignments.
5. Works with appropriate personnel to supervise and monitor registration of all motor vehicles parking on school grounds as well as bus transportation.
6. Conducts periodic inspections of the high school building and grounds and reports needed work and repairs to the appropriate persons.
7. Effectively implements school and district action plans in crisis and emergency situations.

Record Keeping and Financial Management

1. Maintains accurate personnel, student, and fiscal records.
2. Works with the principal to prepare accurate budgets and monitor expenditures.
3. Prepares weekly discipline reports and disseminates information to appropriate school and district personnel.
4. Provides information to the principal for required district reports.
5. Collaborates with the principal to identify future building and equipment needs and to plan appropriate action.

CORNWALL-LEBANON SCHOOL DISTRICT
Job Description: Elementary School Principal

A. Supervision

_____ 1. Supervises and assesses the instructional performance of the professional staff.
_____ 2. Implements an induction program for newly assigned personnel.
_____ 3. Monitors the implementation of the written curriculum.
_____ 4. Emphasizes student achievement and growth as the primary goal of the school.
_____ 5. Encourages faculty to communicate information with parents.

B. Administration

_____ 6. Implements board policy, state laws, and contractual obligations.
_____ 7. Collects and analyzes data, when appropriate, prior to making a decision.
_____ 8. Maintains accurate personnel, pupil, and financial records and provides information as needed.
_____ 9. Requests and allocates available resources in an effective and efficient manner.
_____ 10. Assists in the supervision of the classified staff.

C. Climate

_____ 11. Identifies and recognizes staff contributions within and outside the school community.
_____ 12. Models appropriate attitudes and behaviors to staff and students.
_____ 13. Promotes an aesthetically pleasing environment in the building and on the grounds.
_____ 14. Maintains appropriate communications with district administration.
_____ 15. Maintains high visibility in the school.

D. Communications

_____ 16. Engages in communication accurately, sensitively, and reliably.
_____ 17. Displays good organizational skills and coherence in oral and written communication.
_____ 18. Manages conflict effectively.
_____ 19. Demonstrates the ability to use a variety of group process skills in interaction with staff, parents, and students.

_____ 20. Listens and responds appropriately to staff, student, and community concerns.

E. Student Activities

_____ 21. Maintains appropriate student conduct through the development and implementation of a fair and effective behavior management system.

_____ 22. Oversees the planning and implementation of the student activities program in their building.

_____ 23. Conducts programs and activities to recognize student achievement and promote self-esteem.

_____ 24. Facilitates the development of human relations skills.

_____ 25. Insures the proper educational placement of pupils and seeks appropriate services for students with special needs.

_____ TOTAL

_____ AVERAGE (Total ÷25)

_____   _____
Administrator's Signature   Evaluator's Signature

CORNWALL-LEBANON SCHOOL DISTRICT
Descriptors for Job Performance Criteria
Elementary School Principal

A. Supervision

1. Individually consults with teachers on a plan for observation and supervision. Agrees on which teachers are to be formally evaluated and which will be subject to short visits. Monitors to what degree teachers are using the various instructional strategies.
2. Conducts (on a monthly basis) a meeting with new staff members on a particular subject.
3. Uses a variety of techniques including observation, lesson plans, and exams.
4. Studies student test scores, mid-term, and final exams. Emphasizes to staff the importance of accountability. Analyzes achievement and growth.
5. Asks faculty to keep him/her informed on important information they are communicating to parents.

B. Administration

6. Implements board policy, state laws, and contracts consistently. When in doubt, requests a verification.

7. Does not jump into making a decision without careful planning and discussion with others.
8. Includes attendance records, permanent record cards, discipline files, student activity accounts, health records, report cards, schedules, and eligible student records.
9. Uses budgeted funds in a fair and consistent manner. Uses an effective procedure for teachers wanting to attend conferences.
10. Assists in the evaluation of these employees. Checks on their work performance. Feeds back to the business office areas of concerns.

C. Climate

11. Writes notes to staff members. Alerts newspaper to do a feature story. Suggests that parents write positive notes to teachers.
12. Sets an example through appearance, warmth, sincerity, consistency, concern, fairness, tactfulness, etc.
13. Takes pride in building. Student work is displayed where possible. Grounds are well maintained.
14. Keeps district office informed of matters that are of a critical nature and may require board involvement.
15. Is visible in the corridors, classrooms, and at extracurricular events.

D. Communications

16. Weighs words carefully with students, parents, and staff. Has an ''I care'' attitude.
17. Checks written communication to the public for accuracy and grammar. Communications to parents must be positive. Uses good speech patterns in discussions with students and staff.
18. Is a problem solver and a compromiser. Has the ability to work through a problem to its successful conclusion.
19. Involves others in the discussion. Gives the impression that he or she is genuinely interested in what is being said. Uses brainstorming techniques where appropriate.   Uses visual aids to demonstrate a point.
20. Uses listening as an important skill. Hears the person out. Responds in a confident manner to what is being said.

E. Student Activities

21. Ensures that students feel they are treated fairly. Hears students' concerns. Insists on proper behavior and appropriate dress.
22. Plans activities and places on the school calendar. Insures proper supervision of each activity.
23. Has a variety of activities aimed at student recognition. Encourages activities to promote self-esteem.
24. Encourages staff to incorporate human relations skill building into classroom activities. Develops pride among students in their school.

25. Works closely with counselors and psychologists on pupil placement. Works with various community agencies on proper treatment and assistance for students.

CORNWALL-LEBANON SCHOOL DISTRICT
Interpretation of the Rating Scale

2 – Meritorious

The meritorious rating indicates a degree of performance that can be achieved but rarely maintained. It is a level of performance that can be reached through a performance objective or an unusual situation that occurs. All meritorious ratings must be accompanied by a written statement containing specific examples that demonstrate the nature of the performance.

1 – Satisfactory

The performance of the administrator meets the requirements of the position in a satisfactory manner throughout the rating period. This standard implies that the individual has produced what can reasonably be expected of a fully competent person in this position.

0 – Needs Improvement

The performance of the administrator does not meet all the requirements of the position. This rating indicates more effort and/or understanding between rater and employee or other corrective action. A suggestion on how to improve must accompany this rating.

Examples of Individual Administrative Objectives

The first objective provided below is for an assistant high school principal with primary responsibility for student discipline in his or her school.

OBJECTIVE — The existing student discipline program at the high school will be reviewed, recommendations for improvements will be developed, and the new program will be implemented fully by the beginning of the next school year.

The assistant principal will form a committee of interested teachers to study the current school discipline program. Surveys of student and teacher opinions regarding the positive and negative elements of the present system will be developed, administered, and analyzed by the committee. The assistant principal will provide student suspension and detention data for the previous five years for committee review.

Pertinent research on the topic of effective student discipline programs will be made available to the committee for its review. An implementation plan for a revised student discipline program will be prepared by the assistant principal for consideration by the committee. The revised student discipline program, as approved by the committee, will be submitted to the superintendent prior to the end of the current school year.

The second objective presented is for an elementary principal. It concerns an initiative by the principal to explore the pros and cons of a new method for teaching Language Arts in the elementary school.

OBJECTIVE— *During the current school year the elementary principal will acquire a thorough knowledge and understanding of Whole Language programs for the elementary grades and will prepare a summary report of his or her findings for the school faculty and the superintendent.*

The following activities will be undertaken to meet this objective:

(*1*) The principal will attend two or more regional workshops or conferences on the topic of Whole Language.

(*2*) The principal will access two or more Whole Language video programs via the district's satellite disc communications system.

(*3*) The principal will form a study group of interested staff members to discuss topics relating to Whole Language.

(*4*) The principal will visit at least three nearby schools that have initiated various components of a Whole Language program.

(*5*) The principal will develop a bibliography of research on Whole Language and read and study selected materials based on the bibliography.

(*6*) The principal will prepare a summary report for the superintendent outlining his or her activities and conclusions regarding the topic of Whole Language.

The last objective provided is for a middle school principal. This objective is designed to address the perennial problem of involving parents in the school once their children have reached middle school age.

OBJECTIVE— *The middle school principal will undertake initiatives to develop a proactive, influential, and effective Parent Teacher Organization at Technomic Middle School.*

To achieve this objective, the principal will attempt to recruit able and interested volunteers from the existing parent organization to serve as officers and as chairs of various committees relating to proposed new programs. These potential new initiatives could include, but would not be limited to, the following possibilities:

(*1*) A parent newsletter edited by parents
(*2*) A series of evening programs sponsored by the PTO
(*3*) Parent participation on existing school committees relating to curriculum, school climate, and student discipline
(*4*) PTO sponsorship of student awards for academic and/or extra-curricular achievement
(*5*) PTO sponsorship of a Teacher Recognition Program

A report will be prepared for the superintendent at the conclusion of the year detailing the status of the above suggested initiatives as well as any other proposals that might be initiated by the PTO group.

Suggested Readings

Anderson, M. E. 1989. "Evaluating Principals: Strategies to Assess and Enhance Their Performance," *OSSC Bulletin,* 32(8):62.

Egginton, W. M. 1988. "State Mandated Tests for Principals—A Growing Trend?" *NASSP Bulletin,* 72(502):62-71.

Ginsberg, R. and Barry, B. 1989. "Influencing Superior's Perceptions: The

Fudge Factor in Teacher and Principal Evaluation," *Urban Review,* 21 (1):15-34.

Gomez, J. and R. S. Stephenson. 1987. "Validity of Assessment Center for the Selection of School Level Administrators," *Educational Evaluation and Policy Analysis,* 9(1):1-7.

Gottfredson, G. D. and L. G. Hybl. 1989. "Some Biographical Correlates of Outstanding Performance Among School Principals," Center for Research on Elementary and Middle Schools, Baltimore, MD. 19 pages.

Herman, J. J. 1990. "Evaluating the Performance of School Business Officials," *School Business Affairs,* 56(2):10-14.

Hughes, L. W., M. E. Murphy, and M. J. Wong 1986. "This Administrator Assessment Center Aims for Excellence," *Executive Educator,* 8(3):21-22 + .

Ingram, R. L. 1986. "Strategic Planning Must Be Linked to Performance," *School Administrator,* 43(3):9-11.

Joines, R. C. and J. Hayes. 1986. "Assessment Centers Like This Are the Rage for One Reason: Because They Work," *Executive Educator,* 8(12): 22-23.

Lawton, S. B. 1986. "Development and Use of Performance Appraisal of Certificated Education Staff in Ontario School Boards," Ontario Department of Education, Toronto.

Lindahl, R. A. 1986. "Implementing a New Evaluation System for Principals," *Planning and Changing,* 17(4):224-32.

Manatt, R. P. 1987. "Lessons from a Comprehensive Performance Appraisal Project," *Educational Leadership,* 44(7):8-14.

Murphy, J. 1986. "The Administrative Control of Principals in Effective School Districts: The Supervision and Evaluation Functions," *Urban Review,* 18(3):49-75.

Rist, M. C. 1986. "Principals Mull the Merits of the New Evaluation Techniques," *The Executive Educator,* 8(4):37, 43.

Robinson, G. E. and P. M. Bickers. 1990. "Evaluation of Superintendents and School Boards," *ERS Report,* RIENOV 90.

Snyder, W. R. and W. H. Drummond. 1988. "Florida Identifies Competencies for Principals," *NASSP Bulletin,* 72(512):48-58.

Shilling, J. L. 1986. "Developing an Operational Plan: Maryland's Initiative for Quality Leadership," *NASSP Bulletin,* 70(486):11-14.

Valentine, J. W. and M. L. Bowman. 1989. "Principal Effectiveness in National Recognition Schools, A Research Summary Report" (ERIC Document #ED311552).

CHAPTER FIVE

Performance Appraisal of the Superintendency Team

THE superintendency team includes all district office administrators, including the superintendent. In an extremely small school district of only several hundred students, the superintendent might well be responsible for all of the business, personnel, curriculum, and other functions typically assigned to district-level administrators in larger districts.

For the purposes of this discussion, the authors have assumed that the administrative positions described in this chapter exist in a school district with a student enrollment of 2,000–4,000. Such an enrollment base would typically require a superintendent, a chief financial officer, and two or three other administrators in the areas of personnel, curriculum and instruction, and/or pupil services.

Special Characteristics of District-Level Administrative Positions

These direct subordinates of the superintendent provide him or her with both the breadth and depth of management services necessary to effectively administer and lead a large and complex organization. These administrators are the people that the superintendent must most directly rely upon to carry out his or her own major goals for the school district. District office personnel are also viewed by the staff and community as representing the goals, priorities, and values of the superintendent and the school board.

Clearly the members of the superintendency team are in a position to have a significant impact on the success of the school district and the achievement of specific district goals and superintendent objectives. On a daily basis they will interact most often with building administrators. As mentioned previously, principals are often fully engaged with the challenges of their own school and are often reluctant to devote their energies toward the achievement of district-wide goals and objectives.

Thus members of the superintendency team must operate on the reflected authority of the superintendent in an environment where their efforts are often undervalued by building-level staff. Success in this environment requires an uncommon combination of diplomacy and assertiveness on the part of the district administrator. Such central office personnel also run the risk of becoming so narrowly focused on their own areas of responsibility that they in turn can lose sight of the complexities and nuances that principals and teachers must deal with on a daily basis.

Dedication to the objectives of the superintendent is certainly to be highly valued in a district office administrator. Such personnel should also possess good abilities to collect and analyze data relating to their areas of responsibility. Planning and organization skills should also be of a high order. They should have sufficient people skills so that they will rely primarily on persuasion rather than on formal power to achieve necessary cooperation and collaboration from building administrators and other district-level administrators.

Supervising the District-Level Administrator

The above characteristics of the working environment of district office administrators suggest certain strategies for effective supervision. In evaluating central office personnel, the superintendent will need to place a great deal of reliance upon reports and other documents relating to particular objectives or job duties.

Was the math curriculum completed in a thorough and timely manner? How accurate were the revenue and expenditure projections for the year? Were all positions filled in a timely manner by qualified personnel? Was a tracking system for students referred for special education services developed and successfully implemented? How well was the new staff development program rated by teachers and administrators? Properly crafted reporting documents can provide the superintendent with much meaningful information on the performance of his or her direct subordinates.

The relationships that district office personnel establish with building principals and other district personnel are of great importance in evaluating the effectiveness of district administrators. Data on such relationships can best be gathered indirectly and informally. One of the major benefits of frequent school visits by the superintendent is that such frequent visits will encourage more informal conversations between principals and the superintendent.

These frequent informal conversations will lead to the discussion of topics and perceptions by principals that would not be discussed at formal and infrequent evaluation conferences. Through these discussions the superintendent will over time achieve a good understanding of the working relationships between district office personnel and the building administrators and teachers. Such feedback can be helpful to the superintendent in his or her efforts at coaching immediate subordinates toward better performance.

Another source of feedback on the manner in which the district administrator is being perceived at the building level is an evaluation survey to be completed by each building administrator relative to district office personnel. Although the district office administrators should be encouraged to share the results of these surveys with the superintendent, they should not be required to do so.

It is reasonable to assume that administrators who receive negative feedback from building administrators will attempt to correct the deficiencies, even if the data are not shared with the superintendent. Most district-level administrators would want to improve their ratings from their building-level colleagues and then share these more positive results with the superintendent.

As part of a continuous management and appraisal process, the superintendent should meet weekly with his immediate subordinates to discuss progress toward objectives as well as the status of normal job functions. These meetings, the frequent informal discussions with building principals, the survey results on the effectiveness of district administrators supplied by the principals, and the formal documentation on the achievement of specific objectives will provide the necessary process and data for effectively appraising the performance of district administrators.

Coaching the District-Level Administrator

Several of the factors in the working environment of central office administrators alluded to above create special pitfalls for these administrators and unique challenges for the superintendent as coach. In the paragraphs to follow, the authors will draw on their personal experiences with central office administrators to illustrate coaching and supervisory issues common to these positions.

We will look first at administrators who become so narrowly focused on their own areas of responsibility that they cut themselves off from interactions with any staff members or issues not directly relating to their work. These administrators become the embodiment of the definition of experts as those who know more and more about less and less until they eventually know everything about nothing.

Over the years, the isolation of this person becomes so complete that he or she is unable to recognize, much less identify, the large majority of staff members. This administrator becomes isolated from the overall mission of the organization and unfamiliar with the challenges and objectives of his or her administrative colleagues. In one situation of this sort known to the authors, the problem was corrected over time by the superintendent's insistence that the administrator involve himself actively in policy-setting meetings of all administrators, that the administrator visit the schools frequently to meet with principals concerning his areas of responsibility, and by assigning the administrator to special projects outside his normal areas of responsibility.

Another supervisory challenge more common with central office administrators is motivating the mid-career professional. Most often people are promoted to central office positions after long-term careers as teachers and building administrators. They are often both knowledgeable and conscientious but lack the enthusiasm of the young professional.

It is incumbent on the superintendent to provide these valuable employees with job challenges that will provide motivation. The superintendent must also be the one to recognize and reinforce positive achievements, since the nature of the work of district administrators dictates that their efforts will be unrecognized by their building-level colleagues and will almost certainly go unnoticed by the school board or the community.

These special motivational challenges must play to the strengths of the mid-career administrator. A central office co-worker of one of the authors had a special talent for methodically and tenaciously interpreting the byzantine guidelines necessary to comply with federal and state regulations for special education and Chapter One remedial programs. Capitalizing on these talents, this administrator was able to develop excellent procedures and policies in these areas, which served the school district well and provided enhanced job satisfaction to the administrator.

Another area in which district office administrators often require guidance concerns the image of the central office that they project at the building level. Although district office people would be quick to insist that they exist to serve the teachers and principals at the building level, those on the front lines are inclined to view central office as a bureaucratic maze imposing paperwork and other superfluous requirements, thus interfering with the real work of the schools.

This view of central office by building-level professionals can be especially damaging since all central office personnel are seen as an extension of the superintendent. Thus, even a superintendent who personally tries to avoid imposing burdens on his or her building staffs may have the reputation for bureaucratic interference because of the actions and practices of his or her subordinates.

In supervising our immediate subordinates we have found it useful to consistently remind them that the central office exists to remove barriers and serve the needs of teachers and administrators at the building level. In arriving at Individual Administrator Objectives we always tried to ensure that at least one of these objectives was designed to streamline the delivery of district-level services to the building level. We also used the objectives process to extend the job responsibilities and perspectives of individual central office personnel and to promote a more service-oriented mode of thought at district office.

Evaluating a central office administrator as unsatisfactory presents a special set of challenges for the superintendent. To begin with, the superintendent may be the only person in a position to know that his or her immediate subordinate is not doing the job. This contrasts with the typical situation involving an unsatisfactory teacher or principal where performance deficiencies are often obvious to fellow staff members and even to students and parents.

Lack of competence on the part of a central office administrator is often not detected by others because of the narrow range of the position or the infrequency of meaningful contact between this administrator and other administrators or teachers. Since a lack of satisfactory performance is not obvious to others, it becomes more difficult for the superintendent to confront the administrator and demonstrate his or her areas of incompetence. The task is complicated by the fact that the central office administrator often works closely with the superintendent on a daily basis, thus forming closer personal ties than the superintendent might have with building-level employees.

The above special considerations for supervising district office personnel speak to the critical need for a thorough and well-documented evaluation process. Through judicious use of job descriptions and IAOs, the superintendent will be able to enforce reasonable performance standards on the work of his or her less able subordinates. In a worst case situation, such documentation can help the superintendent to successfully counsel his or her subordinate into retirement or another career.

On the following pages we have provided examples of Job Descriptions and Individual Administrative Objectives for three different district office positions in our fictitious Technomic School District. These positions are Assistant Superintendent for Curriculum and Instruction, Director of Personnel, and Director of Business Affairs. A sample survey of the performance of the Assistant Superintendent for Curriculum and Instruction to be completed by the building administrators has also been included. Finally, an authentic example of Individual Administrative Objectives, as well as a mid-year progress report on these objectives, is provided.

TECHNOMIC SCHOOL DISTRICT
Job Description: Assistant Superintendent
for Curriculum and Instruction

Reporting Relationship

    1. Reports to superintendent of schools.

2. Works closely with principals.
3. Assumes the responsibilities of the superintendent in his or her absence or as delegated.
4. Supervises district curriculum coordinators and the school psychologist.

Primary Functions
1. Assists the superintendent by providing leadership in curriculum and instruction, special education, and staff development.
2. Provides coordination for student services programs.
3. Assists in selection and supervision of teachers.

Performance Responsibilities

Curriculum Development, Supervision, and Evaluation
1. Provides leadership for K-12 curriculum and program development, implementation, and evaluation through working relationships with principals and curriculum coordinators and by implementing the Curriculum Management Plan; chairs the district Curriculum Council.
2. Plans and coordinates all curriculum writing projects, including summer workshops.
3. Insures that the written curriculum is taught and that planned courses meet state requirements.
4. Provides leadership for program evaluation by analyzing both standardized and criterion-referenced test results.
5. Assists in the supervision and evaluation of the instructional staff.
6. Assists principals as requested in determining course selection criteria, teacher assignment, and student grouping for instruction.
7. Provides opportunities and encouragement for staff to increase program expertise.

Special Education
1. Provides leadership for the district's special education program by coordinating curriculum and instruction and all due process procedures, forms, and correspondence, as well as the services provided by the school psychologist.
2. Works with Intermediate Unit special education staff to insure compliance with all state regulations and standards and to strengthen the district program.
3. Handles all decisions and procedures regarding out-of-district placements, insuring that students in these programs are appropriately and properly served.
4. Makes referrals to appropriate community agencies when needed.

Professional Development
1. Serves as the co-chair (with a teacher representative) of the Professional Development Committee, overseeing the implementation of the Professional Development Plan and the Teacher Induction Program.

2. Identifies, plans, and implements, in conjunction with the Professional Development Committee, all district in-service programs in accordance with assessed needs.
3. Keeps abreast of current staff development programs, consultants, and speakers at the local, state, and national levels, identifying appropriate ones for possible district use.
4. Encourages staff members to participate in professional development experiences that will contribute to their personal growth on the job.

Decision Making and Problem Solving
1. Makes decisions in a timely fashion and maximizes decision effectiveness by follow-up actions.
2. Identifies problem areas and seeks solutions before crisis situations develop.
3. Seeks information from appropriate sources, delegates decision making effectively, and strives for consensus.
4. Considers research when making decisions.
5. Clearly communicates decisions and rationale to all affected.

Communications
1. Listens and responds clearly and appropriately to administrator, staff, community, and student concerns.
2. Fosters open communication among staff.
3. Communicates effectively with administrators, staff, school board, and community.
4. Keeps the superintendent and principals informed of district needs and problems.

General Administration
1. Consults with the superintendent and principals in the preparation of the annual budget.
2. Prepares and maintains accurate records and files for federal, state, and special education programs.
3. Prepares clear and effective district reports.
4. Coordinates student services as assigned, e.g., nurses, library.
5. Handles routine administrative matters effectively.

Personal development
1. Strengthens skills and knowledge through participation in professional activities and organizations.
2. Perceives self as a change agent, continually working to improve district programs and staff effectiveness.
3. Demonstrates openness to new ideas, flexibility, decisiveness, loyalty.
4. Supports and promotes district goals.

Community Relationships
1. Interacts with school district and parent groups to promote positive outcomes.

2. Keeps the community informed about school activities through newsletters, attendance at parent meetings, etc.

Record Keeping and Financial Management
1. Maintains accurate personnel, student, and fiscal records.
2. Prepares accurate budgets and effectively monitors expenditures.
3. Prepares required district reports accurately and efficiently.
4. Handles routine administrative matters effectively.
5. Anticipates future building and equipment needs.

TECHNOMIC SCHOOL DISTRICT
Building Administrators Evaluate the Assistant Superintendent for Curriculum and Instruction

Please circle the number that best represents your opinion of the assistant superintendent's performance. Circle 5 if the statement is always true, circle 4 if it's almost always true, circle 3 if it's frequently true, circle 2 if it's occasionally true, circle 1 if it's never true, and circle NA if it's not applicable.

1. Provides leadership for K-12 curriculum and program   1 2 3 4 5 NA
   development, implementation, and evaluation through
   working relationships with principals and curriculum
   coordinators and by implementing the Curriculum
   Management Plan; chairs the district Curriculum
   Council.
2. Provides leadership for program evaluation by analyz-   1 2 3 4 5 NA
   ing both standardized and criterion-referenced test
   results.
3. Assists in the supervision and evaluation of the instruc-   1 2 3 4 5 NA
   tional staff.
4. Assists principals as requested in determining course   1 2 3 4 5 NA
   selection criteria, teacher assignment, and student
   grouping for instruction.
5. Provides leadership for the district's special education   1 2 3 4 5 NA
   program by coordinating curriculum and instruction
   and all due process procedures, forms, and correspondence, as well as the services provided by the school
   psychologist.
6. Handles all decisions and procedures regarding out-of-   1 2 3 4 5 NA
   district placements, insuring that students in these
   programs are appropriately and properly served.

7. Identifies, plans, and implements, in conjunction with the Professional Development Committee, all district in-service programs in accordance with assessed needs. 1 2 3 4 5 NA
8. Encourages staff members to participate in professional development experiences that will contribute to their personal growth on the job. 1 2 3 4 5 NA
9. Makes decisions in a timely fashion and maximizes decision effectiveness by follow-up actions. 1 2 3 4 5 NA
10. Seeks information from appropriate sources, delegates decision making effectively, and strives for consensus. 1 2 3 4 5 NA
11. Listens and responds clearly and appropriately to administrator, staff, community, and student concerns. 1 2 3 4 5 NA
12. Communicates effectively with administrators, staff, school board, and community. 1 2 3 4 5 NA
13. Prepares clear and effective district reports. 1 2 3 4 5 NA
14. Demonstrates openness to new ideas, flexibility, decisiveness, loyalty. 1 2 3 4 5 NA
15. Interacts with school district and parent groups to promote positive outcomes. 1 2 3 4 5 NA

TECHNOMIC SCHOOL DISTRICT
Job Description: Director of Personnel

Reporting Relationship
1. Reports to superintendent of schools.
2. Works closely with all administrative personnel.
3. Directly supervises secretarial staff in the personnel department.

Primary Functions
1. Assists the superintendent and other administrators in the recruitment, selection, orientation, evaluation, and professional development of all school district employees.

Performance Responsibilities

Recruitment, Selection, and Orientation of Personnel
1. Organizes and implements an aggressive program to recruit quality personnel for all school district position categories.
2. Develops personnel selection procedures and practices designed to maximize staff involvement in the identification and selection of new personnel.
3. Collaborates with principals and other administrators to provide effective orientation programs for all categories of new personnel.

Supervision and Evaluation
1. Develops and implements, with appropriate input from staff, a systematic and thorough program for the supervision of all staff members.
2. Develops, with appropriate staff input, forms and procedures for implementing a formal evaluation program for all categories of staff.
3. Monitors current research literature and practices in other school districts to identify possible areas for improvement in the supervision and evaluation process.

Employee Relations Programs and Practices
1. Represents the school district at the bargaining table during collective negotiations.
2. Administers the provisions of collective bargaining agreements and other documents relating to employees' terms and conditions of employment.
3. Represents the school district in ''meet and discuss'' and other meetings with various employee groups for the purpose of developing compensation guides and associated terms of employment.

Staffing Levels and Work Assignments
1. Develops recommendations for the superintendent relating to staffing levels for all categories of staff and programs.
2. Gathers and analyzes comparative data regarding staffing levels and compensation levels from local, state, and national sources.
3. Develops, in conjunction with appropriate personnel, position descriptions for all job assignments within the school district.

Payroll and Benefits Administration
1. Supervises and administers payroll function for all school district employees.
2. Administers medical and other insurance programs for the school district.
3. Periodically researches opportunities to implement more efficient and effective benefits programs.

TECHNOMIC SCHOOL DISTRICT
Job Description: Director of Business Affairs
Reporting Relationship
1. Reports to superintendent of schools.
2. Works closely with principals.
3. Supervises director of operations, food services director, and business office staff.

Primary Functions
1. Assists superintendent by performing basic management functions of planning, organizing, controlling, coordinating, directing, staffing, and evaluating within the areas of school business administration.

Performance Responsibilities

Budget Development and Control
1. Develops and controls a school district operating budget based upon educational needs and priorities identified by the superintendent after consultation with the Superintendency Team.
2. Develops and implements an efficient system for the purchase, distribution, storage, and inventory control of all property, equipment, and supplies.
3. Prepares and submits all required business-related state reports in a timely manner.
4. Prepares monthly financial reports for the school board.
5. Develops and implements a cash management and investment program for the school district.
6. Organizes and implements the tax collection function for the district.

Accounting and Record Keeping
1. Establishes and maintains an accounting system for all funds in conformance with state guidelines and generally accepted accounting principles.
2. Establishes and implements procedures for maintaining all records that prudence and legal requirements demand.
3. Provides the proper safeguards for the custody of public funds.

Operation and Maintenance of School Facilities
1. Directly supervises and evaluates the custodial, maintenance, food services, transportation, and business office operations of the school district.
2. Operates and maintains school facilities so that the educational needs of the district will be met in a safe, comfortable, and efficient manner.
3. Expedites the successful planning and execution of renovation and construction projects by working effectively with architects, attorneys, financial advisors, bidders, and contractors.

Staff Supervision and Evaluation
1. Implements school district procedures for directly supervising and evaluating immediate subordinates.
2. Monitors supervisory program as it relates to food services, transportation, and custodial/maintenance personnel.
3. Participates fully in the selection and orientation programs for new employees.
4. Encourages staff members to participate in training programs that will contribute to their personal growth on the job.

Individual Administrative Objectives for District Personnel

The first example of an IAO provided will be for a director of personnel. This particular objective is intended to correct a perceived problem in the district with respect to the recruitment and selection of new teachers.

OBJECTIVE—The personnel director will conduct a thorough study, review, and subsequent revision of the teacher selection process in the Technomic School District.

Early in the school year the personnel director will form a committee of interested building administrators and subject supervisors to study and recommend revisions in the teacher selection process in the district. This study will include an evaluation of the effectiveness and efficiency of our current practices as well as the development of suitable objectives for a revised system. The committee will then develop a revised set of guidelines to govern the entire teacher selection process.

Existing paper and pencil instruments for use during the interviewing process will be reviewed for the purpose of selecting a suitable tool for adoption in the Technomic School District. All administrators will receive training in the use of the selected instrument during the latter part of the school year. Topics such as advertising for vacancies, paper screening of candidates, procedures for conducting reference checks, and the application of appropriate affirmative action guidelines will also be discussed. A final recommended plan will be presented to the superintendent by May of the current school year.

The role played by the director of business affairs in the success of the management team is often underappreciated. Clearly, the efficient functioning of the business side of the school enterprise maximizes the resources available for the district to fulfill its educational mission. We believe that the chief financial officer or business manager should be considered a vital member of the management team and should participate fully in all aspects of the performance appraisal process.

Next is an example of an objective suitable for a person fulfilling the role of director of business affairs.

OBJECTIVE— During the current school year the director of business affairs will develop a Five Year Capital Equipment Purchase Plan to meet the operation and maintenance requirements of all school district facilities.

The preparation of this Five Year Capital Equipment Purchase Plan will include the following elements:

(1) A report from a consulting architectural firm on the current condition of the mechanical systems in all facilities will be developed along with recommendations for new and replacement equipment to maintain operating efficiencies.

(2) Initial specifications will be developed for any major renovation projects that might be recommended as a result of the survey of existing conditions in the school facilities.

(3) A financial plan will be developed, in consultation with school district auditors and investment bankers, to determine bond issue options available to finance major capital equipment purchases and potential renovation projects.

(4) A comprehensive report including each of the above elements will be presented to the school board for consideration in April of the current school year.

Thus far we have presented examples of Individual Administrative Objectives at both the building and district office levels. Next we would like to offer an example of an actual set of objectives presented to a superintendent at the beginning of a school year. Following this listing is a report by the administrator to the superintendent on the status of the objectives at the time of the mid-year administrative appraisal conference.

The example provided is used by permission of Dr. Karen-Lee Brofee, Director of Curriculum and Instruction for the Octorara Area School District in Atglen, Pennsylvania.

EXAMPLE OF AN ACTUAL SET OF OBJECTIVES

To: Dr. Richard P. McAdams, Superintendent
From: Dr. Karen-Lee Brofee, Director of Curriculum and Instruction
Date: August, 1990
Re: Individual Administrative Objectives for 1990-91

1. To coordinate the implementation, monitoring, and evaluation of the newly revised Mathematics Curriculum, K-12.

The Mathematics Committee that I facilitated through the 89-90 school year, composed of nine teachers of mathematics and three administrators drawn from all three building levels, succeeded in developing an educationally sound, practical, and forward-thinking mathematics curriculum. It is my firm belief that this document can provide the critical support necessary for making a difference in the mathematics literacy of our students if the teachers and administrators work together to prioritize its implementation, monitoring, and evaluation stages. It is appropriate as a district-level individual that I work to structure, support, and reinforce these collaborative efforts among administrator, teacher, and parent. I intend

- to scrutinize all written components for their efficacy, seeking regular feedback from teachers at all levels
- to coordinate the orientation of the written document with all building staffs through the leadership of teachers
- to work with building administrators in their understanding of the curriculum document and the monitoring of its implementation in both formal and informal structures
- to facilitate both the internal and external evaluation of the document through formal and informal hard and soft data and the processing of the gathered material through the curriculum committee's review (student standardized scores on Stanford 8, TELLS, SAT, prognostic testing, teacher grades, student portfolios, student projects, teacher attitude toward the curriculum document, teacher understanding of the document)
- to help communicate the intent and effect of the new curriculum to various publics (administrators, other subject teachers, school board, parents, students)

2. To facilitate the adoption of cooperative learning instructional practices.

   Cooperative learning skill development is currently being overtly pursued by approximately twenty-five teachers across the district and is seen by administrators as holding tremendous potential for increasing student achievement throughout the district. Cooperative learning does require significant training, significant support, and tremendous patience if the academic results are to be realized. As Director of Curriculum and Instruction, I can coordinate curricular efforts to fully utilize cooperative learning as a chosen strategy and facilitate continued training as well as reinforcement for cooperative instructional efforts. I intend

   - to lead an initial awareness and informational workshop on the opening in-service day for approximately twenty teachers
   - to facilitate a Cooperative Learning Team that will provide interested teachers with continued motivation to attempt cooperative learning, to support their efforts through collegial sharing, and to provide a variety of continuing growth opportunities

- to provide a menu of activities and opportunities for teachers still exploring their levels of interest and commitment to cooperative learning, including after-school study sessions, visitation within the district, visitation outside of the district, team-teaching cooperative learning lessons, development of a cooperative learning notebook, and coordinating the cooperative learning workshop through the SERC — Satellite Interactive Staff Development Series on Instructional Strategies for Improving Schools
- to structure self-evaluation processes for teachers in assessing their awareness and efforts
- to collaborate and assume instructional responsibilities with one or more teachers in the cooperative instruction of students
- to collaborate with one or more teachers in the collection of comparative cooperative learning effectiveness data

3. To interpret, analyze, and disseminate the standardized test results of student performance for student, counselor, teacher, and administrator application.

Thorough and clear test analysis can serve students, teachers, and administrators well in planning for maximum academic growth. Following the selection of a particularly rigorous standardized test by the 1989-90 test review committee and an adjustment to late spring testing, it is critical for teachers to realize the practical use of the information for their students' benefit. It is my intent

- to complete a data base entry of student Stanford and TELLS scores enabling the generation of particular classes for teacher and counselor use
- to prepare an analysis of the data by building, grade, and subject area for building principals' use in targeting instructional goals
- to prepare specific reports for curriculum committees' uses
- to prepare a report for public release that will clarify Octorara's measured performance
- to prepare a report for the school board of directors that highlights general strengths and areas for growth

4. To facilitate the revision of the Science, Health, Guidance, and Physical Education Curricula.

During the 1990-91 school year, the District Science Committee will continue its review of the effectiveness of the current curricular and instructional practices K-12, research current trends and issues in science education, and develop a written document for 1991-92 implementation. Separate committees of guidance counselors and health and physical education teachers will be formed to begin the research phase of curriculum revision, culminating in the development of a completed document for implementation in 1992-93.

MID-YEAR STATUS REPORT

To: Dr. Richard P. McAdams, Superintendent
From: Dr. Karen-Lee Brofee, Director of Curriculum and Instruction
Date: December 1990
Re: Mid-Year Conference

Appraisal Item I: Goals and Objectives for 1990-91

1. To coordinate the implementation, monitoring, and evaluation of the newly revised mathematics curriculum, K-12.

   The Mathematics Committee successfully revised, published, and distributed the curriculum. Staff were oriented to it either through peers who served on the committee or by the principal. Evaluations of the orientation to the document were very positive. In my work to date to reinforce the actual application of this document to the classrooms, I have

   - made visitations to classrooms with follow-up notes
   - copied articles and materials to teachers
   - participated in the Mathematics Satellite broadcast and discussed the program with three middle school teachers with follow-up notes to these teachers and to their principals noting their attendance
   - held two discussion sessions with each of the two levels of teachers involved in the rewriting of the curriculum
   - attended the Eastern Mathematics Conference in New Jersey
   - reviewed mathematics grades for 1st Quarter in the high school
   - conferenced with regular classroom teachers and gifted teacher on areas of concern
   - served as consultant to department on mathematical and experimental piloting
   - used math curriculum and math issues in small group work with middle school teachers with PCRP II mini-workshops (Grade 7 and 6)
   - promoted attention to Math performance on SAT, NEDT, and TELLS
   - informally raised awareness of increasing technology in mathematics in middle and high schools
   - created a list of assessment/monitoring questions for principal use
   - collected data for TITLE II audit, which served also to give feedback on the curriculum

   Results:

   - Teachers are seeking me out to share student responses and to discuss issues.
   - The language of ''new'' concepts is being used by teachers.

- Teachers (Elem) are gaining confidence in their use of concrete objects and teaching for true concept attainment.
- The math curriculum is serving as a useful model for other curriculum efforts.

Needs include

- to visit more classrooms more frequently
- to discuss some of the more important pieces with building principals
- to develop an evaluation survey for teachers on the curriculum; use the next committee meeting to consider its use
- to increase teacher awareness of TELLS
- to coordinate the Prognostic Testing through Penn State

2. To facilitate the adoption of cooperative learning instructional practices, I have

- led an initial awareness and informational workshop on the opening in-service day for approximately twenty teachers
- led a skill development workshop for twelve middle school teachers at the October in-service day
- facilitated the development of a Cooperative Learning Team that is providing fifteen interested teachers and administrators with continued motivation to promote and use cooperative learning, to enrich their efforts through collegial sharing, and to provide a variety of continuing growth opportunities including the leadership skills to disseminate cooperative learning among their peers
- led a workshop session with the Cooperative Learning Team
- provided a menu of activities and opportunities for teachers still exploring their level of interest and commitment to cooperative learning; these include after-school study sessions, visitation within the district, visitation outside of the district, team-teaching cooperative learning lessons, development of a cooperative learning notebook, and coordinating the new cooperative learning workshop through the SERC—Satellite Interactive Staff Development Series on Instructional Strategies for Improving Schools
- collaborated and/or assumed instructional responsibilities with five teachers in the cooperative instruction of students (Home Ec, Social Studies, Gifted, Science, and English) and I have one or two more scheduled for January
- used cooperative learning strategies in my curriculum committee work and my mini-sessions on PCRP II
- promoted cooperative learning as a major instructional strategy for the new professional development plan
- evaluated each session that I have led, finding very positive responses

- planned to co-lead the next session of the team with a teacher

Needs include

- to structure self-evaluation processes for teachers in assessing their awareness and efforts
- to facilitate the offering of a graduate-level class on campus in advanced cooperative learning

3. To interpret, analyze, and disseminate the standardized test results of student performance for student, counselor, teacher, and administrator application.

   I have made several efforts this first semester to help teachers and principals to realize the practical use of the information for their students' benefit. I have

   - completed a data base entry of student Stanford and TELLS scores enabling the generation of particular classes for teacher and counselor use
   - distributed classlists to teachers throughout the entire district with SAT scores sorted as they requested for current students
   - met with the 6th and 9th grade teachers to interpret the Writing Assessment results
   - prepared several analyses of the data by building, grade, and subject area for building principals' use in targeting instructional goals and offered to prepare reports of their request
   - prepared specific reports for curriculum committees' uses (specifically SCIENCE)
   - prepared a report for public release on TELLS that will clarify Octorara's measured performance
   - prepared a report for the school board of directors that highlights general strengths and areas for growth
   - prepared a report for high school departments on the performance of the students on the NEDT

4. To facilitate the revision of the Science, Health, Guidance, and Physical Education Curricula.

   I have facilitated the District Science Committee in its continued review of the effectiveness of the current curricular and instructional practices K-12 and research of current trends and issues in science education. Specifically, I have

   - led two full-day sessions of the district committee
   - held formal conversations with individual teachers of articles and material they have selected to read from a list of research articles
   - tabulated and analyzed a survey of student attitude towards science instruction in the high school and the middle school and processed the findings with the teachers

- participated in a field trip to the Franklin Institute
- surveyed surrounding districts concerning their sequence of science courses

Separate committees of guidance counselors and health and physical education teachers have been formed to begin the research phase of curriculum revision.

CHAPTER SIX

Performance Appraisal of the Superintendent

THE superintendent must play a critical role in his or her own evaluation. This is essential since the superintendent's immediate supervisors, the school board members, often have little direct knowledge of the duties of the superintendency. In addition, many board members have no personal experience with the process of fairly evaluating the performance of a subordinate.

Structuring School Board Evaluation of the Superintendent

The appraisal process itself must be structured and implemented in a manner that will serve as an ongoing in-service program on the relative roles and responsibilities of the superintendent and the school board. Specific evaluation procedures must be more detailed than is necessary for the evaluation of other levels of the administrative team. This is required since no previous background or expertise in performance appraisal can be assumed on the part of the school board. In addition, the frequency of school board member turnover virtually assures that in any given year one or more members of the school board will be experiencing the evaluation process for the first time.

Board members are at a disadvantage in evaluating the performance of their superintendent because their major opportunity for direct observation comes at school board meetings. Relationship with the school board and performance at public meetings, although vitally important, represent but a small part of the total duties of the superintendency. Other than this direct observation, board members must rely on secondhand observations and impressions of the superintendent as reported to them by parents, community members, and school staff.

Any superintendent who achieves anything of significance will also accumulate a small number of vocal critics. These opinionated taxpayers, usually unencumbered by any knowledge of the nature of school administration, will complain at school board meetings and through letters to the editor of the local newspaper about the incompetence and obtuseness of the school superintendent. Although these attacks are usually unfounded and often irrelevant, they can distract the school board from attending to other important aspects of the superintendent's performance.

The occurrence of a major controversy in a district, with its attendant attacks on the school by the community, can for many months be the predominant context in which the school board views the performance of its superintendent. Controversies surrounding land acquisition, school construction, teacher strikes, and sex education curricula are just a few of the issues that can distort the context in which the school board evaluates the superintendent. This situation can occur just at the time that the superintendent is most in need of the support and understanding of his or her school board.

Another possible circumstance for which a systematic superintendent evaluation program can provide at least a partial remedy is the case of the board member with the proverbial axe to grind. Sometimes there may be two or three board members who are inclined to base their evaluations on isolated incidents or on one or two decisions by the superintendent. Since the dynamics of an entire board evaluating the superintendent almost always includes a discussion of the superintendent's performance without the superintendent present, the opportunity for the few to unduly influence the many is readily apparent.

Thus it is incumbent upon the superintendent to ensure that a thorough, balanced, and straightforward process is employed to formally evaluate the superintendent. The absence of such a process virtually guarantees that the tenure, evaluation, and compensation of the superintendent will be decided by whim and by chance rather than by the use of a systematic, reliable, and valid process.

Such a lack of an effective evaluation system was evident in a school district known to the authors where a school board first praised its superintendent and shortly thereafter renewed his contract for five years. These events occurred immediately after the superintendent had successfully represented the school board in achieving a long-term teacher contract within parameters set by the board. Within a few months the school board was so disenchanted with this same superintendent that they bought out his contract for $250,000. Clearly, the superintendent evaluation system in this district was in critical disarray.

The process should be structured in a manner that will provide the school board with the maximum amount of data covering as broad a spectrum of the superintendent's duties as possible. By contract the superintendent should be guaranteed at least one meeting with the school board per year during which time his or her evaluation will be the sole matter of discussion.

Documentation for Superintendent Appraisal by the School Board

The principal document in the evaluation process should be the job description of the superintendent. Individual board members should evaluate and rate the level of performance of the superintendent in each major category. Through discussion of this document with board members, the superintendent can periodically reacquaint school board members with his or her critical job responsibilities.

The second major element in the evaluation process should be a report by the superintendent on the status of his or her Individual Administrative Objectives for the year. These objectives will have been jointly developed by the superintendent and the school board at the beginning of the evaluation cycle. Every effort should be made by the superintendent to provide as much supporting information as possible relative to the achievement of these objectives.

As practicing superintendents, the authors made periodic reports on the progress of one or more of their objectives at school board meetings throughout the school year. They also publicly reviewed their objectives with the school board at the beginning of the school year. In this manner they were able to provide both the board and the community with detailed information on each objective in small, digestible doses. The summary report to the board at the end of the year, though rather comprehensive, nevertheless dealt with information that had been discussed with board members at an earlier meeting. Thus, at the evaluation meeting itself, it was not necessary to spend an inordinate amount of time reviewing the IAOs.

The superintendent should have conducted a survey of the administrative staff's perceptions of his or her performance as a part of the data-gathering process. An example of a survey form to accomplish this is included at the end of this chapter. The results of this survey should be shared with the school board. The school board as a whole should review all of this information and discuss questions or concerns regarding it with the superintendent. They should also have an opportunity to share their perceptions with one another without the superintendent present.

A critical part of the process is the completion of a self-evaluation by the superintendent. The superintendent should rate him/herself on all major elements of the job description and should also provide a detailed report on the achievement of his or her Individual Administrative Objectives. This information should be available to school board members before they begin their discussion of the superintendent's performance.

Relating the Superintendent's Evaluation to Compensation

Once all of the above has occurred, the school board members should independently complete the evaluation documents, including numerical ratings. These ratings should be forwarded to the school board president for tallying and preparation of a summary of the school board's evaluation for the superintendent.

The relationship between evaluation and compensation should be clearly stated in the superintendent's employment contract. (A sample of such a contract is included at the end of this chapter.) The school board rating should automatically generate compensation adjustments without the need for further discussion by the school board.

This feature of automatic salary adjustments based on a formula is especially critical at the superintendency level. In all but the most affluent communities, the superintendent will be among the highest paid residents of the community. It is not uncommon for the superintendent to have a higher income than each of the board members who must approve salary increases. Critical board members will find it particularly difficult to properly compensate their superintendent unless salary criteria and adjustment formulas are clearly enumerated in the superintendent's contract.

On the following pages we have provided samples of school board policies, guidelines, procedures, and forms relating to the employment and evaluation of the superintendent. An example of a form for evaluation of the superintendent's performance by members of the administrative team is also included. Several examples of specific Individual Administrative Objectives suitable for a superintendent have also been provided.

TECHNOMIC SCHOOL DISTRICT
School Board Policy
Employment of Superintendent

The position of superintendent is established by law, as are details of election, term of appointment, minimum salary, and general duties. The superintendent is considered a ''commissioned officer'' rather than a professional employee of the district.

The law gives the superintendent a seat on the school board and the right to speak on all matters before the board, but specifies that he or she shall have no vote.

Appointment of a superintendent is a function of the board. It may seek the advice and counsel of other individuals or an advisory committee, and it may hire consultants to assist in selection. However, final selection shall rest with the board after a thorough consideration of qualified applicants. The appointment shall require a majority vote of full board membership and meet other requirements of state law.

The superintendent's appointment shall be secured through a written contract that shall state the term of the contract, compensation, benefits, and other conditions of employment.

State law requires that a district superintendent be appointed for a term of from three to five years from the first day of July next following election or from a time mutually agreed upon by the board and the superintendent. Action must be taken by the board at least 150 days prior to the expiration of the superintendent's term to determine whether the board will re-elect the superintendent or consider other candidates for the position.

Within the legal requirements of the state, the superintendent shall be responsible for the general management of the schools of the district. The superintendent is responsible for guiding the development of the educational objectives and program of the school district to fulfill the educational needs of all pupils and to recommend these objectives and programs to the board. The superintendent shall provide overall direction to the activities of the school district and its personnel in the accomplishment of approved educational goals, administer the policies of the board, conserve the district's assets and resources, and maintain and enhance the district's standing in all relationships.

The management responsibilities of the superintendent shall extend to all activities of the district, to all phases of the educational program, to all parts of the physical plant, and to the conduct of such other duties as may be assigned by the board.

The superintendent shall

- manage the work of all professional and nonprofessional personnel in planning and program development and shall direct all activities of the district. The superintendent may delegate these responsibilities, together with appropriate authority, but he or she may not delegate nor relinquish the ultimate responsibility for the results or for his or her accountability.
- manage the development of long-range and short-range educational objectives for the improvement and growth of the school district and of the educational activities of the district
- manage the development of the overall educational process and administrative procedures and controls necessary to the implementation of educational programs
- manage the regular and systematic evaluation, analysis, and appraisal of the achievements of pupils and the performance of personnel in each of the educational programs
- report to the board the progress and status of programs and activities of the district and inform the board on all matters of major importance or significance in relation to these programs and activities
- establish and maintain an administrative organization that provides for the effective management of all the essential functions of the school district
- recommend revisions to the organization of the management structure, including the establishment or elimination or revision of administrative positions
- develop and recommend policies and programs for recruitment, selection and employment, employee relations, benefits and services, employee safety, evaluation and salary administration
- ensure the maintenance of an adequate staff of properly trained administrative and supervisory personnel throughout the district
- recommend to the board the selection, employment, assignment, transfer, and suspension of professional and nonprofessional personnel
- supervise assigned personnel and conduct periodic evaluations and appraisals of their performance

- recommend salary increases and adjustments for professional and non-professional employees
- develop and recommend to the board job classifications for all new positions
- direct the development of the annual budget for the district
- review and recommend programs and supporting data for funds to be included in the annual budget
- provide for the overall management of the district's financial activities and take appropriate action to assure that expenses are kept within the approved budgetary limits of the district
- assist principals and directors in maintaining economy and efficiency in the operation of administrative units
- maintain an active contact with all local, state, federal, and philanthropic programs that could provide financial assistance to the district
- act as executive officer for the board
- act as professional advisor to the board
- attend meetings of the board, with the right to comment on all issues
- prepare the agenda for all meetings of the board and deliver the agenda with pertinent information on each item in advance of the meeting
- participate in the affairs of local, state, and national professional organizations and serve as a representative of the school system and the community at meetings of such organizations
- maintain a cooperative working relationship between the schools and the community and its agencies
- establish and maintain such other relationships within and outside of the district as are required to carry out the responsibilities and duties of the position

TECHNOMIC SCHOOL DISTRICT
School Board Policy
Evaluation of Superintendent

Purpose

Regular, periodic evaluation of the superintendent's performance is a school board responsibility. In carrying out this responsibility, it is recognized that the superintendent is entitled to such a review in an objective and straightforward fashion so that his or her leadership may be as effective as possible for the school district.

Guidelines

Through the evaluation of the superintendent, the board shall strive to accomplish the following:

- clarify the superintendent's role in the school system, as seen by the board, for the benefit of the superintendent
- clarify the superintendent's job description and the immediate priorities of responsibility, as agreed by the board and the superintendent, for the benefit of the board members
- develop harmonious working relationships between the board and the superintendent
- provide administrative leadership of unquestionable excellence for the school district

In line with policy, the board shall periodically develop with the superintendent a set of individual administrative objectives based on the needs of the school system. The superintendent's performance shall be reviewed in accordance with these specified objectives. Additional objectives shall be established at intervals agreed upon with the superintendent.

TECHNOMIC SCHOOL DISTRICT
Guidelines
Evaluation of Superintendent

The Technomic School Board wants to be assured that its superintendent is effective in providing leadership for its school system. The superintendent wants to be assured that he or she is doing a good job — indeed, living up to the expectations of the school board and community. This evaluation system provides that assurance for both the school board and superintendent.

The Technomic system of evaluation contains the following essential features:

- Periodically, the superintendent and school board identify needs or areas to emphasize in the coming year. Needs emerge from the district's mission, goals, special problems or projects, or aspects of the superintendency that need strengthening.
- In December, the superintendent and school board establish specific individual administrative objectives and action plans.
- Monthly, the superintendent provides the school board with IAO progress reports.
- By November 1, the superintendent provides the school board with a self-assessment indicating the extent to which his or her objectives and performance standards were achieved.

- By November 1, all members of the administrative team rate the superintendent. A summary of their assessment is forwarded to the board president and superintendent.
- By mid-November, individual board members independently rate the superintendent's performance on the objectives and in six major areas of the job description:

  Curriculum and Instruction Management
  Staff Personnel Management
  School Board Relations
  School-Community Relations
  Financial Management
  Personal Qualities

- By late November, the board president convenes the board members to discuss their assessments and to prepare a summary that best describes the superintendent's performance. Appropriate recommendations for the coming year are included. (Performance Standards account for 50 percent of the evaluation and Individual Administrative Objectives account for 50 percent.)
- By December 1, a copy of the consensus evaluation is transmitted to the superintendent. This includes an overall rating of superior, good, fair, or unsatisfactory.
- During the first week of December, a conference is scheduled with the superintendent, board president, and vice president to discuss the evaluation. Follow-up plans are formulated and discussed.

TECHNOMIC SCHOOL DISTRICT
Procedures and Forms
School Board Evaluates the Superintendent

To: School Board Members
From: Board President
Re: Superintendent's Evaluation

We will be meeting in executive session on Saturday, November _____, at 8:00 A.M., in the Board Conference Room, to evaluate our superintendent. It is extremely important that you read all of the material in this packet as soon as possible. You will need to fill out the three rating forms prior to our executive session. Please bring the complete packet of information to that meeting.

Our superintendent has prepared considerable information to help us in the evaluation of his (or her) performance. This information is attached to the rating forms.

Form #1 — Individual Administrative Objectives

After reading the superintendent's reports, write a narrative assessment for *each individual administrative objective*. Use the key to determine how many points the superintendent deserves *for each objective*. Add the five scores to determine the total points. Then, divide the total points by the number of objectives to determine the average numerical rating in this category.

Form #2 — Performance Standards

After reading the superintendent's self-assessment and the composite rating of the superintendent done by the administrative team, write a narrative assessment for *each major category*. Add the six scores to determine the total points. Then, divide the total points by six to determine the average numerical rating in this category.

Form #3 — Summary of Ratings from Individual Board Members

Transfer the ratings from Forms 1 and 2 to this form. Add the average numerical ratings from the two forms to determine the total points. Using the key, determine the superintendent's overall rating.

TECHNOMIC SCHOOL DISTRICT
Form #1 — Superintendent's Individual Administrative Objectives

Scale

Rating	Points
Superior	3
Good	2
Fair	1
Unsatisfactory	0

1. Explore evaluation and curricula processes developed in the last five years to determine how they are presently used and how we can use them more effectively. These include teacher and administrator evaluation processes, the use of curricula guides and criterion-referenced tests, and strategies to help the marginal performer.

_____

_____

_____

_____

_____

Points_____

2. Recommend an administration staffing plan based on enrollment trends, our organizational structure, future administrative needs, and present administrative team personnel.

_____

_____

_____

_____

_____

_____

Points_____

3. Explore the instructional and financial consequences of specific alternative educational delivery systems, particularly those that utilize technological advances.

_____

_____

_____

_____

_____

_____

Points_____

4. Study student adjustment to ninth grade and formulate specific approaches to improve transition to high school, if necessary.

_____

_____

_____

_____

_____

_____

Points_____

5. Seek long- and short-term ways of reducing costs without jeopardizing the quality of our education program.

_____

_____

_____

_____

_____

_____

Points_____

Total Points for Individual Administrative Objectives   _____

Divide total points above by number of objectives (five) to determine the average numerical rating in this category.   [   ]

TECHNOMIC SCHOOL DISTRICT
Form #2 — Superintendent's Performance Standards

Scale

Rating	Points
Superior	3
Good	2
Fair	1
Unsatisfactory	0

Curriculum and Instruction Management
1. Understands all aspects of the instructional program.
2. Requires school programs to reflect sound, research-based educational practices.
3. Monitors the effectiveness of the instructional program.
4. Shows alertness to new knowledge that might benefit students or faculty.

_____

_____

_____

_____

Points_____

Staff Personnel Management
5. Involves others in the decision-making process while maintaining responsibility for the final decision.
6. Exhibits openness and compassion in dealing with others.
7. Develops good staff morale.
8. Uses evaluation instruments for performance, training, and promotion of staff.

_____

_____

_____

_____

Points_____

Board Relations
9. Keeps board fully informed about operations in district.
10. Gains confidence of board members.
11. Encourages free flow of information from administrators to board committees.
12. Supports board actions and implements them at the best level possible.

_____

_____

_____

_____

Points_____

School-Community Relations
13. Responds to problems and opinions of all groups and individuals.
14. Interprets the educational program to the community.
15. Handles news media relations skillfully.
16. Gains respect and support of the community.

_____
_____
_____
_____

Points_____

Financial Management
17. Determines the educational needs of the district.
18. Plans for and allocates resources fairly and effectively.
19. Prepares a realistic budget and keeps spending within budget.

_____
_____
_____
_____

Points_____

Personal Qualities
20. Exhibits imagination and competence in planning, organizing, and follow-through.
21. Attempts to maintain an objective view when solving problems.
22. Collects adequate information before making decisions. Does not delay important decisions nor allow pressure to cause hasty decisions.
23. Maintains composure in handling self in a variety of situations.
24. Handles a variety of complex issues at one time.
25. Communicates clearly and thoroughly.
26. Exhibits integrity in all dealings.

_____
_____
_____
_____

Points_____

Total Points for Performance Standards   _____

Divide total points above by six to determine the average
numerical rating in this category.

TECHNOMIC SCHOOL DISTRICT
Form #3—Superintendent's Evaluation
Summary of Ratings from Individual Board Members

Superintendent's Individual   Average Numerical Rating_____
Administrative Objectives

Superintendent's Performance Standards   Average Numerical Rating_____

Total Points   _____

Divide total points above by two to determine superintendent's
overall rating.   _____

Scale for Overall Rating

Points	Rating
2.7 - 3.0	Superior
1.8 - 2.6	Good
1.0 - 1.7	Fair
less than 1.0	Unsatisfactory

TECHNOMIC SCHOOL DISTRICT
The Administrative Team Evaluates the Superintendent

Scale

Rating	Points
Superior	3
Good	2
Fair	1
Unsatisfactory	0

Performance Standards

Points   **Curriculum and Instruction Management**

_____   1. Understands all aspects of the instructional program.

_____   2. Requires school programs to reflect sound, research-based educational practices.

_____   3. Monitors the effectiveness of the instructional program.

_____   4. Shows alertness to new knowledge that might benefit students or faculty.

Staff Personnel Management

_____   5. Involves others in the decision-making process while maintaining responsibility for the final decision.

_____   6. Exhibits openness and compassion in dealing with others.

_____ 7. Develops good staff morale.
_____ 8. Uses evaluation instruments for performance, training, and promotion of staff.

Board Relations
_____ 9. Keeps board fully informed about operations in district.
_____ 10. Gains confidence of board members.
_____ 11. Encourages free flow of information from administrators to board committees.
_____ 12. Supports board actions and implements them at the best level possible.

School-Community Relations
_____ 13. Responds to problems and opinions of all groups and individuals.
_____ 14. Interprets the educational program to the community.
_____ 15. Handles news media relations skillfully.
_____ 16. Gains respect and support of the community.

Financial Management
_____ 17. Determines the educational needs of the district.
_____ 18. Plans for and allocates resources fairly and effectively.
_____ 19. Prepares a realistic budget and keeps spending within the budget.

Personal Qualities
_____ 20. Exhibits imagination and competence in planning, organizing, and follow-through.
_____ 21. Attempts to maintain an objective view when solving problems.
_____ 22. Collects adequate information before making decisions. Does not delay important decisions nor allow pressure to cause hasty decisions.
_____ 23. Maintains composure in handling self in a variety of situations.
_____ 24. Handles a variety of complex issues at any one time.
_____ 25. Communicates clearly and thoroughly.
_____ 26. Exhibits integrity in all dealings.

Comments

_____
_____
_____
_____
_____
_____
_____
_____

Total Points for Performance Standards _____

Divide total points above by twenty-six to determine average rating.
┌─────────────┐
│   │
└─────────────┘

═══

═══

TECHNOMIC SCHOOL DISTRICT
Superintendent Employment Contract

This agreement made the _____ day of (*month/year*) by and between Technomic School District, a school district organized under the laws of the State of _____ having its principal place of business at (*address*) _____
_____ (the "District")

A
N
D

Dr. (*name of supt.*) _____, residing in (*County*) _____,
(*State*) _____ (the "Superintendent")

WITNESSETH:

1. The School Board of the Technomic School District (the "Board") by resolution adopted on the _____ day of (*month*) in accordance with (*state*) law elected Dr. (*name of superintendent*) as its Superintendent of Schools for the term of five (5) years commencing July 1, (*year*) and terminating June 30, (*year*), and authorized the execution of this Employment Contract with the Superintendent.

2. The Superintendent represents that he or she holds all certificates and credentials required by (*state*) law and desired by the District to accept this position.

3. The Superintendent agrees to relocate his or her residence to the District. Upon completion of such relocation the District will pay a relocation allowance to Superintendent in the sum of $(*amount*).

4. The Superintendent agrees to perform the duties of Superintendent of Schools in a competent and professional manner subject to the established policies and regulations of the Board and the laws of (*state*).

5. The School District agrees to pay to the Superintendent for the first year of the term covered by this Contract a salary of $(*amount*), payable bi-monthly according to a daily rate of $(*amount*), for total working days per year of two hundred sixty (260). Annual increases to the initial salary

shall take place during the remaining years covered by this Contract. Such increases shall take into consideration the merit of the performance by the Superintendent of his or her duties and obligations hereunder as determined by the Board and shall consider, *inter alia*, an annual performance review thereof by the Board.

6. A portion or all of at least one meeting annually of the Board shall be devoted to a private discussion of the working relationships between the Board and Superintendent, evaluation of job performance, and discussion of goals for the ensuing year.

7. In addition to the duties prescribed by law, the following shall be the duties and obligations of the Superintendent:

   a. The Superintendent will furnish recommendations to the Board on all matters having to do with selection, appointment, assignment, transfer, promotion, organization, reorganization, reduction, or termination of personnel employed or to be employed by the School District, all subject to final approval by the Board.

   b. Administration of the affairs of the School District, including but not limited to programs, personnel, and business management, will be lodged with Superintendent, and all duties and responsibilities therein will be performed and discharged by the Superintendent or by his or her staff under his or her direction.

   c. The Superintendent shall have a seat at the Board table and the right to speak (but not vote) on all issues before the Board in accordance with applicable law. The Superintendent and/or his or her designee(s) shall attend all regular and special meetings of the Board and may attend committee meetings thereof, and will serve as advisor to the Board and said committees on all matters affecting the School District.

   d. Criticism, complaints, and suggestions called to the attention of the Board or its members will be referred to the Superintendent for study, disposition, or recommendation as appropriate.

8. In addition to the foregoing salary, the Superintendent shall receive all of the benefits set forth in the document attached hereto marked Exhibit ''A'' and made a part hereof.

9. The School District further agrees that it shall to the full extent permitted by law, defend, hold harmless, and indemnify the Superintendent from any and all demands, claims, suits, actions, and legal proceedings brought against the Superintendent either in his or her individual capacity or in his or her official capacity as agent and employee of the School District, provided the incident arose while the Superintendent was acting or, in good faith, reasonably believed himself or herself to be acting within the scope of his or her employment as Superintendent of the School District.

10. It is mutually understood and agreed that this Contract may be modified only by mutual agreement of the parties, and all such modifications and agreements shall be evidenced by written and executed amendments to the Contract. Such amendments shall be subject to the same provisions of law as this original Contract.

11. Except as required by law, no right to receive payments under this Contract shall be subject to anticipation, commutation, alienation, sale, assignment, encumbrance, charge, pledge, or hypothecation, or to execution, attachment, levy, or similar process or assignment by operation of law, and any attempt, voluntary or involuntary, to effect any such action shall be null, void, and of no effect.
12. The District may withhold from any compensation or benefits payable hereunder all federal, state, local, and other taxes as shall be required pursuant to any law, governmental regulation, or ruling.
13. No term or condition of this Contract shall be deemed to have been waived nor shall there be any estoppel against the enforcement of any provisions of this Contract except by written instrument of the party charged with waiver or estoppel.

IN WITNESS WHEREOF, the parties hereto have affixed their hands and seals this _____ day of (*month/year*).

ATTEST: TECHNOMIC SCHOOL DISTRICT

By _____   By _____
              Secretary   President, School Board

WITNESS: SUPERINTENDENT

_____   _____ (SEAL)
                                                  (name)

Exhibit "A" to Superintendent's Employment Contract
Additional Benefits

1. The Superintendent shall receive all benefits including, but not limited to, medical insurance, hospitalization, including major medical, dental care, prescription drugs, vision care, life insurance, disability leave, academic reimbursement, and physical examination as are provided to administrators under the District's Guidelines for Compensation of Administrators ("Administrators' Compensation Guidelines") as now in force and as hereafter amended, from time to time, during the term of this Contract.
2. The Superintendent shall receive the following additional benefits and in the event any said benefits at any time conflict with benefits provided

under the Administrators' Compensation Guidelines, the Superintendent shall receive the benefits most favorable to him or her:

a. The Superintendent shall be entitled to twenty (20) vacation days per year to be taken at the discretion of the Superintendent after reasonable advance notice thereof to the President of the Board. The Superintendent will be compensated at the per diem rate for all unused and accumulated vacation days upon termination of employment with the School District.

b. The Superintendent shall be entitled to full reimbursement for tuition for graduate study with prior approval by the Board.

c. The full expense for the Annual Physical Examination provided under the Guidelines shall be paid by the District.

d. At the commencement of employment under this Contract with the District, the Superintendent shall receive credit for 150 days of unused sick leave. Sick leave thereafter shall accumulate as provided in the Administrators' Guidelines.

e. The School District shall pay the Superintendent's membership dues for professional organizations with prior approval by the Board.

f. Income protection for disability will be provided to the Superintendent under the same terms and conditions as applicable from time to time for all Administrators with the exception that the maximum benefit payment per month for the Superintendent shall be $(*amount*).

3. The Superintendent shall be reimbursed by the School District at the rate of $(*amount*) per mile for use of his or her automobile on School District business or at the mileage deduction rate as published periodically by the IRS for tax deduction purposes, whichever is higher.

4. The Superintendent shall be reimbursed by the School District for all other appropriate expenses which are incurred by the Superintendent for School District purposes or business.

5. The Superintendent shall be permitted to attend appropriate education-related conferences and all reasonable expenses shall be reimbursed or paid for by the School District with prior approval by the Board.

6. Notwithstanding any of the aforementioned provisions, the Superintendent shall be entitled to receive all fringe benefits not classified above which are included or expanded for other School District administrators during the term of this Contract.

Suggested Readings

Aplin, N. D. and J. C. Daresh. 1985. "The Superintendent as Educational Leader," *Planning and Changing*, 14(4):209-218.

Bennis, W. and B. Nanus. 1985. *Leaders: The Strategies for Taking Charge.* NY: Harper & Row.

Bippus, S. L. 1985. "A Full, Fair, and Formal Evaluation Will Enable Your Superintendent to Excel," *American School Board Journal,* 172(4):42–43.

Blumberg, A. 1985. *The School Superintendent: Living with Conflict.* NY: Teachers College Press.

Booth, R. R. and R. Glaub. 1978. *Planned Appraisal of the Superintendent.* Illinois Association of School Boards, Springfield.

California School Boards Association. 1985. *Evaluating Your Superintendent.* Sacramento, CA: California School Boards Association.

Calzi, F. and R. W. Heller. 1989. "Make Evaluation the Key to Your Superintendent's Success," *American School Board Journal,* 176(4):33–34.

Cuban, L. 1985. "Conflict and Leadership in the Superintendency," *Phi Delta Kappan,* 67:28–30.

Educational Research Service, Inc. 1978. *Evaluating Superintendents and School Boards.* Arlington, VA: Educational Research Service, Inc.

Glaub, G. R. 1983. "Board and Superintendent Share Appraisal Benefits," *Updating School Board Policies,* 14(4):7 pages.

Hayden, J. G. 1986. "Crisis at the Helm: Superintendents and School Boards in Conflict," *The School Administrator,* 43(10):17–19.

Hoyle, J. R., F. English, and B. Steffy. 1985. *Skills for Successful School Leaders.* Arlington, VA: American Association of School Administrators.

Konnert, M. W. and J. J. Augenstein. 1990. *The Superintendency in the Nineties: What Superintendents and Board Members Need to Know.* Lancaster, PA: Technomic Publishing Co., Inc.

Lagana, J. F. 1989. "Ready, Set, Empower! Superintendents Can Sow the Seeds for Growth," *The School Administrator,* 46(1):20–22.

Murphy, J. and P. Hallinger. 1986. "The Superintendent as Instructional Leader: Findings from Effective School Districts," *The Journal of Educational Administration,* 24(2):212–231.

Redfern, G. B. 1980. *Evaluating the Superintendent.* Arlington, VA: American Association of School Administrators.

Rogers, J. J. and L. A. Safer. 1990. "Case Study of an Innovative Superintendent Succession Plan," *The Clearing House,* 64(Nov./Dec.):136–140.

CHAPTER SEVEN

Performance Appraisal of the School Board

THE final step in a complete appraisal process for the administrative team is the evaluation of the school board itself. Here again the superintendent has a critical role to play in structuring the form and content of the exercise. A formal school board appraisal program should be mandated by policy and should be implemented through a regularly scheduled annual review.

The minimal time available to school boards to conduct complex activities such as self-evaluation presents a major challenge for public school governance. In reality, a typical school board may have as little as ten to twenty hours per year to discuss such critical matters as policy formation, superintendent evaluation, and school board self-evaluation.

A public school board generally operates in a far different environment than a board of control in the private sector. By law, school boards must consider and approve even minor expenditures and administrative actions, which in the private sector would receive perfunctory attention at best. Also, school boards are political bodies that must respond directly to the concerns of the general public. Thirdly, school boards must conduct virtually all of their business in public.

In the course of a year a typical school board might meet as a full board twice each month for two or three hours. Thus within the space of twelve months this board will have a total meeting time of, at most, seventy-two hours. At least one-third of this time will be spent on routine items such as approval of bills, personnel actions, and reports to the board regarding school district operation.

Another third of the available time will be spent responding to topical matters of concern to the community. Examples of such matters might be taxes, sex education, redistricting, negotiations with employee groups, school building programs, busing concerns, school discipline, etc. These are the problems that produce the newspaper headlines and generate calls to board members and petitions at public meetings. They can also easily distract the school board and the superintendent from attending to the critical areas of superintendent and school board evaluation.

Thus the superintendent must ensure that such limited time that is available for school board evaluation is well planned and well utilized. The documents and procedures discussed in this chapter are designed to produce maximum benefits in the context of the minimal time and effort that can be devoted to the task.

The natural human proclivity to defer analyzing one's own performance, coupled with the normal press of urgent business to be considered by the board, can easily lead to the permanent deferral of the board evaluation process. This can be prevented if the superintendent simply initiates the process each year in accordance with previously approved school board policy. The documents required by the process should be distributed to individual board members and the board president should schedule a definitive time for board members to discuss the results generated by the forms completed by individual board members. The results of the individual evaluations should be submitted to the board president prior to the evaluation meeting so that they can be tallied and summarized.

The evaluation form to be used should be organized in terms of the major responsibilities of the school board and should allow board members to rate the school board as a whole in each responsibility area. The evaluation form should state the board responsibilities in behavioral terms so that the board member is actually rating the extent to which school board behaviors coincide with school board responsibilities. "Building Better Boardsmanship," the following example of such a form, is an evaluation instrument that has been used by 100 school boards in New Jersey and Pennsylvania. It was developed by Anita S. Bieler and one of the authors of this book. The seventy-item instrument is divided into ten performance categories. The seventy criteria were determined after an exhaustive search of the literature on school board effectiveness.

TECHNOMIC SCHOOL DISTRICT
"Building Better Boardsmanship"

Using the following scale (1 = Never; 2 = Sometimes; 3 = Frequently; 4 = Always; 5 = Don't Know) to indicate the degree of existence of the characteristics or traits of the total school board. Circle the most appropriate response for each statement.

1 = Never   2 = Sometimes   3 = Frequently   4 = Always   5 = Don't Know

A.   Board Development

Our school board

1. Orients new members to the nature of the board's duties, policies, and operating procedures.   1 2 3 4 5
2. Reads publications produced by professional education and school board organizations, to keep abreast of trends in education.   1 2 3 4 5
3. Participates in state, regional, and national association functions to become informed of new issues in education.   1 2 3 4 5
4. Designates funds for school board training as an annual budgetary item.   1 2 3 4 5

B.   Meeting of the Board

Our school board

5. Receives the agenda and background information with sufficient time to prepare for the meeting.   1 2 3 4 5
6. Refrains from adding items to the agenda at the last minute.   1 2 3 4 5
7. Comes to the meeting having reviewed the agenda and prepared to conduct a productive meeting.   1 2 3 4 5
8. Has an agenda that reflects the district's priorities in that the most important issues are acted upon first.   1 2 3 4 5
9. Makes a conscious effort to welcome the public by providing agendas and other related materials.   1 2 3 4 5
10. Starts its meetings on time.   1 2 3 4 5
11. Has a president who is in control of the meeting.   1 2 3 4 5
12. Speaks loudly and clearly.   1 2 3 4 5
13. Displays good listening skills, allowing all sides to be heard before a decision is made.   1 2 3 4 5
14. Listens to recommendations prepared by the administration prior to making decisions.   1 2 3 4 5
15. Directs questions about operating procedures to the superintendent.   1 2 3 4 5

1 = Never	2 = Sometimes	3 = Frequently	4 = Always	5 = Don't Know

16. Remains calm under pressure.   1 2 3 4 5
17. Respects differences of opinions and beliefs.   1 2 3 4 5
18. Displays a spirit of compromise and cooperation.   1 2 3 4 5
19. Supports majority decisions.   1 2 3 4 5
20. Holds new, controversial, or complicated issues for discussion only and places them on the next agenda for action, providing preparation time for self and superintendent.   1 2 3 4 5
21. Refrains from abuse of the privileges of executive sessions, tabling issues, committee meetings, and special meetings.   1 2 3 4 5
22. Periodically reviews the meetings to assess their accomplishments and future needs.   1 2 3 4 5

C. Fiscal Management

Our school board
23. Participates in long-range financial planning.   1 2 3 4 5
24. Maintains a financial reserve of at least 5 percent of its operating budget.   1 2 3 4 5
25. Budgets according to the total needs of the system.   1 2 3 4 5
26. Assists the community in understanding the budget.   1 2 3 4 5
27. Seeks available outside funds, such as state and federal allocations and grants.   1 2 3 4 5

D. School-Community Relations and Communications

Our school board
28. Seeks cooperation with various news media.   1 2 3 4 5
29. Communicates systematically with employees.   1 2 3 4 5
30. Communicates systematically with residents.   1 2 3 4 5
31. Makes deliberate advance communications efforts on major policy issues through surveys, advisory committees, or public hearings.   1 2 3 4 5
32. Designates representatives of the board to meet with other local governing bodies and community groups to discuss matters of mutual concern.   1 2 3 4 5
33. Adheres to channels of communication through the superintendent for concerns, complaints, and criticisms.   1 2 3 4 5
34. Makes facilities and resources available to the community.   1 2 3 4 5
35. Plans communication in the event of crisis.   1 2 3 4 5

1 = Never	2 = Sometimes	3 = Frequently	4 = Always	5 = Don't Know

E. Relationship with the Superintendent

Our school board
36. Provides the superintendent with a clear statement of its expectations.   1 2 3 4 5
37. Communicates with the superintendent within the spirit of mutual trust and confidence.   1 2 3 4 5
38. Provides time for the superintendent to plan.   1 2 3 4 5
39. Requests information from staff members through the superintendent.   1 2 3 4 5
40. Discusses potential problems between and among the board and administrators at the earliest opportunity.   1 2 3 4 5
41. Provides opportunity for the professional growth of the superintendent.   1 2 3 4 5
42. Provides for an annual evaluation of the superintendent.   1 2 3 4 5

F. Instructional Management

Our school board
43. Designates times at board meetings for curriculum presentations.   1 2 3 4 5
44. Reviews program requirements and approves course changes to improve the curriculum.   1 2 3 4 5
45. Encourages the participation of the students, professional staff, and community in the development of the curriculum.   1 2 3 4 5
46. Limits the influence of special interest groups on the curriculum.   1 2 3 4 5
47. Requires follow-up studies of the district's graduates.   1 2 3 4 5
48. Visits the schools to see the physical plant and observe the educational process.   1 2 3 4 5

G. Planning and Goal Setting

Our school board
49. Involves students, parents, teachers, and administrators in goal setting.   1 2 3 4 5
50. Consults with community groups, service organizations, local governing bodies, the state department of education, and others.   1 2 3 4 5
51. Reviews the district's progress toward meeting its goals and revises them as necessary.   1 2 3 4 5

1 = Never	2 = Sometimes	3 = Frequently	4 = Always	5 = Don't Know

52. Establishes new goals based on its own evaluation, the evaluation of the superintendent, and the evaluations of the administrative staff.   1 2 3 4 5

H. Staff Relations

Our school board
53. Holds most employees in high esteem.   1 2 3 4 5
54. Requires clearly defined job descriptions.   1 2 3 4 5
55. Encourages professional growth and increased competency through staff development, in-service programs, visitations, and conferences.   1 2 3 4 5
56. Adheres to a well-defined plan for staff evaluations.   1 2 3 4 5
57. Reserves adequate management rights in labor relations agreements.   1 2 3 4 5

I. Legislative Leadership

Our school board
58. Meets with area legislators to discuss state and/or federal education legislation.   1 2 3 4 5
59. Informs the state school boards association of legislative priorities.   1 2 3 4 5
60. Takes public positions on pending state and federal education legislation.   1 2 3 4 5
61. Includes a legislative report on the board agenda.   1 2 3 4 5

J. Policy-making

Our school board
62. Operates according to its written policies.   1 2 3 4 5
63. Reviews and updates policies and regulations.   1 2 3 4 5
64. Involves administrators, teachers, students, parents, and community members in the development of policy.   1 2 3 4 5
65. Makes available copies of its policies and regulations to students, teachers, parents, and the public.   1 2 3 4 5
66. Adheres to the role of policymaker, leaving the administrative function to the superintendent and staff.   1 2 3 4 5
67. Selects officers on the basis of ability.   1 2 3 4 5
68. Works as a group, not as individuals.   1 2 3 4 5
69. Represents the total public interest in decisions, not self-interest.   1 2 3 4 5
70. Protects the administrative team from unjust criticism.   1 2 3 4 5

Since school board members are drawn from the ranks of the general public, it is quite common that at least some school board members will have little concept of the specific duties of school board membership. An evaluation program with a formal rating system can redirect the attention of board members toward their duties on a regular basis. Equally important, the evaluation discussion will provide school board members and the superintendent with an opportunity to review and critique the operation of the school board as a corporate body.

These discussions will provide an opportunity to clarify the perennial confusion between policy and administration, between the role of the school board and the role of the superintendent. It is helpful to conduct this discussion as part of a general review of roles rather than as part of a debate about the resolution of a particularly thorny issue.

The purpose of this evaluation process is to reinforce areas of good performance by the board and to identify areas where further development is necessary. Time constraints on school boards will make it impractical to further review board performance on a regular basis as a regular part of school board operations. Thus, the further development of the board on areas of concern can best be achieved through in-service programs provided by the state and national school board associations and/or individualized development programs delivered by consultants.

A simple yet powerful in-service exercise would be for the school board to complete its regular evaluation form in terms of ideal practice as well as actual practice. These two documents could then be compared to identify areas of board performance where the actual practice varies the most greatly from the ideal. These specific areas could then serve as topics for future school board development.

Goal setting is an essential element in the effective operation of a school board. The results of the school board's evaluation of its own operation can naturally generate one or more goals for future development by the school board. There also should be a close correspondence between the goals of the superintendent and those of the school board. In many cases the superintendent's goals will be a product of discussions between the board and the superintendent regarding priorities for the school district.

Throughout this book the authors have strongly recommended that Individual Administrative Objectives be included as a critical part of the evaluation process. While it is certainly important that school boards have goals and objectives, care must be taken that school board objectives be both appropriate and achievable.

A school board must be careful that the goals it adopts are sufficiently global that they can provide a general direction for the school district. A school board needs to be thinking in terms of long-range plans, strategic planning, and mission as compared with specific objectives to attain more limited goals.

A school board that adopts objectives that require implementation at the operational level increases the likelihood that it will step beyond policy development into the area of administration. One area in which a school board could appropriately establish a specific goal is in matters concerning its own operation. Appropriate topics might include budget development, in-service training, self-evaluation, superintendent/board relationships, and community relations.

Below are a few examples of school board objectives relating to several of these areas:

(1) The school board will engage a consultant to conduct a school board workshop to clarify the respective roles of the superintendent and the school board in policy development and administration.

(2) Individual school board members will participate actively in building-level and district-level long-range planning committees and will keep the total school board informed regarding the priorities and goals of the various individual committees.

(3) School board members will individually and collectively study and discuss various approaches to school budget development and will approve a new school board policy relating to the budget development process.

The school board should be careful to keep the number of objectives chosen for one year to a maximum of two. This limitation is necessary because of the limited time available for such activities as discussed previously. The superintendent will need to be responsible for ensuring that opportunities are available for the school board to actively pursue the one or two objectives that are selected.

Earlier in this chapter we mentioned that the self-evaluation instrument, "Building Better Boardsmanship," has been used during the early 1990s by over 100 school boards in New Jersey and Pennsylvania. Most of that work has been for research purposes. Specifically, we have been attempting to determine if there are significant differences between school boards recognized as *effective* and other boards *randomly selected*. The following results of our research to date should prove instructive to school boards wanting to improve their effectiveness.

Board Development

Effective boards consistently score higher than *randomly selected* boards in all areas of board development. Since the largest percentage of board members we've studied have been serving for less than four years, the issue of orientation is of great concern. Veteran board members need to focus their attention on improving their skills by learning more about the changing roles, responsibilities, procedures, and policies of school boards.

Written policies and budgeted monies supporting the orientation and continuing development of school board members through workshops, seminars, meetings, and retreats are more characteristic of *effective* boards. Such boards orient their new members to the nature of the board's duties, policies, and operating procedures more consistently than do *randomly selected* boards. A large amount of information is needed by new board members, much of which can be learned from the veterans on the board. However, when the majority of board members are new and inexperienced, the task of orientation becomes time-consuming and burdensome for the veteran board members. In our research, we find more veteran members on *effective* boards.

Whether by their own initiative or through help from the superintendency team, *effective* boards are more aware of opportunities to improve their boardsmanship skills and are more active in professional associations than are *randomly selected* boards.

Meeting of the Board

According to our research, the average scores of the *effective* boards were higher than the *randomly selected* boards on all eighteen criteria in this category. *Effective* boards do a much better job in preparing for board meetings. These boards are more likely to receive their agendas and background materials with a realistic number of items and sufficient time to read and prepare for the meeting than are the *randomly selected* boards. Overall, *effective* boards appear to be better prepared for their meetings. Being prepared is of course essential for the conduct of a good board meeting.

Effective boards elect their officers on the basis of ability. They have presidents who are in control of the meeting and who provide leadership for the board. Such boards follow accepted parliamentary procedures more often than do *randomly selected* boards. They adopt specific procedures and are guided by their presidents to stay within them.

Effective boards are more likely than *randomly selected* boards to adhere to their prescribed role as policymaker, leaving the administrative functions to their administrators. In an attempt to remain within their policy-making functions, these boards listen to recommendations prepared by their superintendents and direct subsequent questions and comments to their superintendents prior to making decisions. If new, controversial, or complicated issues that were not on the agenda are brought up, they are discussed briefly and placed on the next agenda for action. In so doing, these more able boards are able to provide preparation time for themselves and their superintendents.

Effective boards are more likely than *randomly selected* boards to share the discussion time so that all board members can be heard. They listen to all sides and ask questions to clarify issues. Members of such boards display a spirit of compromise and cooperation; they respect differences of opinions more often than do *randomly selected* boards. If board members respect other opinions, they are more likely to support majority decisions even when the decisions are contrary to their own. *Effective* boards work as groups, not as individuals.

Effective boards start their meetings on time. They also try to end meetings within two to three hours. This forces board members to stay on task and resolve issues in a timely manner. These boards welcome the public to their meetings by providing agendas, minutes, and related materials. They locate their meetings in settings that are convenient and comfortable for conducting public business.

Such boards conduct their meetings in the spirit of the "Sunshine Laws" and refrain from abusing the privileges of executive sessions, tabling issues, committee meetings, and special public meetings more often than do *randomly selected* boards.

Effective boards review their meetings to assess their accomplishments and future needs more frequently than do *randomly selected* boards. These highly skilled boards are more likely to evaluate their progress and learn from their mistakes — an internal method of board development. *Effective* boards perceive themselves as being prompt, available, and open to the public. They emphasize the need for timely agenda preparation, smooth operation of meetings, attention paid to the public, and cooperation among members. In all cases, *effective* boards score higher in these criteria than do *randomly selected* boards. The meeting is the public appearance of the board. It is here that the public makes its judgments of the quality of its board.

Fiscal Management

Effective boards budget according to the total need of the system more often than do *randomly selected* boards. They are more likely to participate in long-range financial planning and maintain an adequate financial reserve. They budget for proper maintenance of their buildings and equipment, require up-to-date valuations of their property, and provide adequate insurance coverage.

Effective boards assist the community in understanding the school district budget more often than do *randomly selected* boards. This is a time-consuming activity; it involves understanding the intricate details of the budget. Boards must thoroughly understand the budget process, they must format the budgets to be understood by their constituents, and they must have the courage to stand before the public to defend their proposed expenditures.

School-Community Relations and Communications

Effective boards communicate more systematically with employees than do *randomly selected* boards. In doing so they are able to channel the communication of concerns, complaints, and criticisms through their superintendents and thus protect their administrators from unjust criticism. They also provide more in-service training in public relations for all staff members.

Effective boards utilize regular, effective two-way communication systems with the public more frequently than do *randomly selected* boards. They make more deliberate advance communications efforts through surveys, advisory committees, or public hearings on major policy decisions and plan more communications in the event of crises.

Such boards are more likely to seek cooperation with the various news media. Boards need to realize that they are subject to media scrutiny and, therefore, should develop procedures for dealing with it. High performance boards handle those responsibilities more skillfully and perceive themselves as having better relationships with the media.

Effective boards see the need to involve the community in their decision making. They are more likely than the *randomly selected* boards to appoint citizens' committees to advise and help solve specific problems, solicit suggestions from the community when filling vacant board positions, and designate board representatives to meet with local governing bodies and community groups. They also give greater support and cooperation to PTA and other citizens' groups and encourage community attendance at public functions. By involving the community, the boards are seeking suggestions from their public prior to making decisions rather than hearing criticism after the decisions are made.

Effective boards make their facilities and resources available to the community more often than do *randomly selected* boards. The opening of computer labs, libraries, gyms, auditoriums, and athletic fields to the general public is one way to make the schools accessible to all segments of the community, not just the students. These boards are more likely to evaluate their district public relations program than are *randomly selected* boards.

Relationship with the Superintendent

Effective boards provide their superintendents with clear statements of their expectations. They evaluate their superintendents more regularly than do the *randomly selected* boards. In evaluating the superintendent, *effective* boards are also evaluating the effectiveness of other school personnel. The most competent school boards want effective superintendents who are interested in having effective schools.

Effective boards communicate with their superintendents within the spirit of mutual trust and confidence more frequently than do *randomly selected* boards. They support the management team concept of school district operation. They have closer working relationships with their superintendents and feel that trust is the key strength in those relationships. As a result, *effective* boards feel they can discuss potential problems with their administrators at the earliest opportunity. In contrast, *randomly selected* boards are less willing to intervene quickly into potential board/administrative problems. Perceptions of effectiveness can be hindered by the lack of frank, timely discussions.

Effective boards are more likely to offer their superintendents professional-level salaries. Although both groups see the importance of offering a professional salary to acquire and retain competent personnel, the *randomly selected* boards are less likely to base the superintendent's salary on performance evaluation than are the *effective* boards.

Highly skilled boards request information from staff members through the superintendent's office more often than do *randomly selected* boards. While this is true, *effective* boards do not follow the prescribed "chain of command" as often as they feel they could. Board members need to be careful not to use their positions on the board to require staff members to respond to their unauthorized requests, thus neglecting their primary responsibilities.

The superintendent is a non-voting member of the board. It is, therefore, necessary for the relationship between this "CEO" and the "board of directors" to work harmoniously and productively for the good of the students and the staff. The superintendent must be an integral part of board development activities so that he or she, too, can contribute to the board's productivity.

Instructional Management

Effective boards are more likely to make curricular and co-curricular decisions on the basis of what is best for all students than are *randomly selected* boards. They are also more conscious of providing equal access to all activities for all students.

Such boards designate time at their board meetings for the discussion of the curriculum more regularly than do other school boards. *Effective* boards encourage the participation of the professional staff, students, and community in the development of the curriculum more often than do *randomly selected* boards. They also make more attempts to limit the influence of special interest groups on the curriculum. These boards do a better job of reviewing program requirements, approving course additions and deletions, and making the necessary improvements in the curriculum more often than is typically the case.

According to our research, *randomly selected* boards visit the schools to observe the physical plant and the educational process more often than do *effective* boards. Visitations to the schools by board members have been a source of much discussion among educators. Teachers and principals often question how a school board can make decisions about schools without observing them firsthand. Our findings are consistent with the literature, which indicates that *effective* boards are less involved with the operation of their schools than are *randomly selected* boards. They are more likely to assign these responsibilities to their administrators, who are hired to oversee the schools. Many boards do not become involved in the development of curriculum. They see that as a function of the professional staff. However, they feel it is their responsibility to assess the outcomes.

Planning and Goal Setting

Much to our surprise, *randomly selected* boards are more likely to involve administrators, teachers, students, and parents in the development, review, and revision of goals than are the *effective* boards. However, *effective* boards consult with service organizations, community groups, the state department of education, local governing bodies, and others more often than do other school boards. In addition, *effective* boards establish new goals based on the evaluation of their superintendents and administrative staff more often than do *randomly selected* boards.

Staff Relations

Effective boards see themselves adhering to a well-defined plan for staff evaluations. They hold most employees in high esteem, they approve job descriptions, and they have more clearly defined employee competency standards than do *randomly selected* boards. Such boards are more apt to encourage professional growth and increased competency of personnel at all levels through staff development, in-service programs, visitations, and conferences.

These boards pay adequate salaries to entice and retain well-qualified teachers and administrators in their school district more often than do *randomly selected* boards. Conscious of the implications of collective bargaining, *effective* boards are more likely to negotiate clear, exact, and easy-to-implement contracts. From our experience, we know that boards who understand the negotiating process usually maintain the management prerogatives necessary to run their school districts.

Legislative Leadership

Effective boards include legislative reports on their board agendas more often than do *randomly selected* boards. They are also more likely to take public positions and meet with area legislators about pending state and federal legislation.

Effective boards are represented on regional, state, and/or national legislative committees more often than are *randomly selected* boards. These boards familiarize themselves with state and federal groups and organizations concerned with education more often than other boards. They inform the state school boards association of legislative proposals and resolutions. However, the active involvement in school boards associations does not appear to be a priority.

Policy-making

More *randomly selected* boards have revised and updated policies, maintained well-defined policy manuals, and encouraged public comment on policy before adoption than have the *effective* boards. Our findings are contrary to others which indicate that *effective* boards are more inclined to include the public in decision-making activities. However, *effective* boards involve administrators, teachers, students, parents, and community members in the development of policy more often than do *randomly selected* boards.

The more competent boards are more likely to operate according to their written policies than other boards. They seek policy recommendations from their superintendents more often than do *randomly selected* boards. They follow a more corporate structure of listening to the recommendations of their chief executive officer, the superintendent.

Once adopted, copies of policies, rules, and regulations from the *effective* boards are distributed to teachers, students, parents, and the public. These boards utilize a plan for feedback on policies to determine their effectiveness or the need for change more often than *randomly selected* boards.

Mindful of his roles as school board member and facilitator of the board's budget committee and aware of the propensity of some of his colleagues to engage in micro-management, Thomas McKearn, Vice President of the New Hope-Solebury School Board (PA) had this to say after using the "Building Better Boardsmanship" instrument:

> I would like to make a few remarks at this the outset of the series of meetings that will constitute the budget process for our school district.
>
> We all know that a budget deliberation is a difficult and long journey. It can be said that the process has brought out both the best and the worst in our school board. I want to say to you my fellow board members that the basic task before us is not simply one of juggling numbers. The basic interplay is *not* one of the board vs. the administration or of old board members vs. new board members. Rather, the basic issue is what kind of a future we as a board are trying to build for this community. We are part way through a multi-year curriculum reform program and a long-range plan for the district. Programs of this sort cannot be built in a day or a year, but unfortunately, a petulant or overly timid school board can, by denying support on one fateful evening, set back programs of the district by several years.

As school board directors, we each took an oath of office when we began our terms on this board, and last year this board passed a Code of Ethics to provide guidelines for our individual and collective conduct during debates such as the one which we are about to enter. In so doing, we have pledged ourselves to the furtherance of education in our community and in our Commonwealth. Therefore, there can be no disagreement amongst the members of this board as to our obligation for the future of this district and within that general framework we individually and collectively have acknowledged the primacy of providing the best education possible for each of the children in our district.

Reasonable people can disagree over the future which they would like to see for their school. I would suggest that if we are to make this budget process a coherent and efficient process, we would do well to address the major strategic issues of program existence and scope before diving into tactical issues of how such programs might be implemented. In other words, we should not conclude that a program is over- or under-funded without receiving some explanation of what that program is intended to do, and further, if the program is underway, whether its progress to date has been satisfactory.

We must come to view the future not as a threat, but as an opportunity. As duly elected officials we have responsibilities to all our citizenry. However, our primary mission as school board directors must be to ensure that the children entrusted to our care graduate from this school with the educational skills which will enable them to compete in the 21st century. Our task is *not* to define the minimum program of education which will allow board members to pass the "Red Face Test" of meeting some minimal literacy or learning standards. Rather, our obligation is to define and support the best educational program with the maximum achievable goals which this community can support. The process of defining and supporting such a program is what this budget process is all about.

So, in sum, the budget process is the way this board will confront the future. It will be the clearest statement to our community of what we believe in and where each of us stands. For the sake of that process, and for the sake of the children whose futures depend on this process, I would pray that we could, in the next several weeks, develop a collective wisdom and commitment which transcends any partisan feeling which we each might feel. In the end, a simple majority of the board will rule.

The above discussion of the research findings regarding the actual behavior of *effective* boards versus *randomly selected* boards clearly demonstrates the usefulness of the "Building Better Boardsmanship" instrument in identifying the characteristics of an effective school board. The authors recommend the use of this tool as an easy, yet powerful, approach to evaluating school board performance.

Suggested Readings

Bippus, S. L. 1985. "Leap Obstacles to Board Leadership with This Simple Evaluation Process," *American School Board Journal,* 172(9):46-47, 51.

Campbell, R. F., L. L. Cummingham, R. Nystrand, and M. D. Usdan. 1985. *The Organization and Control of American Schools,* Fifth Edition. Columbus, OH: Charles E. Merrill Publishing Co., Chap. 8, pp. 167-194.

Cunningham, L. L. 1985. "The School Board," *Excellence in Education.* John Mangieri, ed., Texas Christian University Press.

Davies, D. R. 1989. *Road to Board Success: Evaluation . . . Development . . . Improvement.* Alexandria, VA: National School Boards Association, Educational Policies Service (ERIC Document ED312777).

Freeman, J. L., K. E. Underwood, and J. C. Fortune. 1991. "What Boards Value," *The American School Board Journal,* 178(1):32-36 +.

George, J. R. and W. B. Seabrook. 1986. "This Board Benefits by Comparing Community Evaluations with Its Own," *American School Board Journal,* 172(1):38-40.

Goldhammer, K. 1964. *The School Board.* New York: The Center for Applied Research in Education, Inc.

Griffith, R. 1990. "When Board Relations Are Smoth, Success Is a Breeze," *Executive Educator,* 12(5):14-16.

Gross, N. 1958. *Who Runs Our Schools?* New York: John Wiley & Sons, Inc.

Kowalski, T. J. 1981. "Here's a Plan for Evaluating Your Board," *American School Board Journal,* 682(7):23.

Kowalski, T. J. 1981. "Why Your Board Needs Self-Evaluation, "*American School Board Journal,* 682(7):21-22.

McCormick, K. 1985. "Here's the Score on How Your Board Can Work in Unison," *American School Board Journal,* 172(6):27-29.

O'Reilly, R. C. 1981. *Some Expectations for New School Board Members.* Dallas, TX: Annual Meeting of the National School Boards Association (ERIC Document ED202160).

Usdan, M. D. 1975. "The Future Viability of the School Board," *Understanding School Boards,* Peter J. Cistone, ed., MA: Lexington Books.

Wellborn, B. 1986. "Tonight's Assignment for All New Board Members: Do Your Homework," *American School Board Journal,* 173(1):37.

Wirt, F. and M. Kirst. 1989. *The Politics of Education: Schools in Conflict.* McCuthan Publishing Coporation.

Zakariya, S. B. 1985. "Well Orchestrated Board Meetings Make Sweet Harmony for Your School," *American School Board Journal,* 172(6):23–27.

———. 1986. *School Boards: Strengthening Grass Roots Leadership.* Washington, D.C.: The Institute for Educational Leadership.

CHAPTER EIGHT

Relating Performance Appraisal to Administrative Compensation

ONE of the more difficult and perennial problems in public school administration is the continuing need to justify salaries paid to school administrators. When considering the school administrator, a large segment of the public sees a typical government bureaucrat operating within a narrow range of policy and allowable practice.

The salaries of these "bureaucratic functionaries" are often far in excess of salaries paid to what are perceived as similar positions in state and local government. Justification of administrative salaries is further complicated by the fact that many people incorrectly assume that school administrators, like most teachers, have a three-month summer vacation.

Public demands for greater accountability in the awarding of administrative salaries are bound to increase as administrative salaries inexorably rise in tandem with significant increases in teacher salaries. Recent teacher contracts extending into the middle 1990s often feature maximum teacher salaries in the $70,000-80,000 range. The number of steps in the salary range are at the same time being compacted so that teachers with as little as twelve to fifteen years of teaching experience will be earning maximum salaries.

What salary level will be necessary to attract an excellent thirty-five-year-old teacher to an entry-level administrative post if his or her earnings as a teacher are in the $70,000-80,000 range? What will be the salary for a starting assistant principal on a twelve-month contract if he or she can earn $75,000 on a nine-month contract? The day of the $100,000 a year building administrator and $150,000 superintendent is rapidly approaching.

An effective administrative evaluation and compensation program can address the issues raised above in addition to fulfilling its primary purpose of improving administrative performance. A systematic compensation program, tied directly to administrative performance, can serve two purposes. The documentation available through the appraisal process can be cited to demonstrate the critical role that administrators play in determining the quality of the schools. A merit compensation plan also provides assurance that both the taxpayer and the administrators are being dealt with fairly by the compensation program.

The current state of school administrator compensation practices in public schools provides a rich opportunity for war stories about inappropriate and counterproductive procedures. Some of these poor practices encountered at various times by the authors include

(1) Tying administrative salaries directly to teacher salary increase. This practice generally eliminates financial incentives for good performance and sometimes leads to a conflict of interests between the teachers and the administrators.

(2) Using impressionistic data from individual board members to determine salary increases for individual administrators. This process breeds favoritism, lowers morale, and undercuts the proper functioning of the superintendent as the immediate supervisor of all other administrators.

(3) Determining administrative increases according to funds available in the budget after all other priorities are addressed. This approach eliminates the possibility of a planned program based on performance and is demoralizing in the extreme to those subjected to it.

(4) Basing compensation decisions on the latest and/or the loudest complaints about an administrator from community, parents, or teachers. This approach robs administrators of a healthy sense of security and control over their own destinies and tends to make them almost neurotically concerned about making controversial decisions or offending anyone. This is not a good frame of mind for someone who must exercise true leadership.

The compensation program should be directly related to the evaluation system as a matter of school board policy. By policy, if not by state law, administrative personnel should have a major role in developing the program by which they are to be evaluated and compensated. A formal mechanism should be developed for establishing salary ranges for each position.

A few sample mechanisms for establishing salary ranges are as follows:

(1) Salary ranges are determined with reference to similar positions in the region or selected nearby school districts.

(2) A job analysis for each position is completed to evaluate the relative knowledge, skill level, job conditions, and responsibility for each administrative position within the organization.

(3) Positions are analyzed with reference to positions in the corporate world and dollar values are assigned to each position relative to the salary range for similar positions in the business world.

An important element of any system selected is that a formula or other method be identified to automatically generate new salary ranges each year to reflect changing market conditions. This practice will make it unnecessary for the school board to grapple with the knotty issue of establishing administrative salaries on a yearly basis.

The administrative salary program should be administered by the superintendent with no direct involvement from the school board. The school budget should contain sufficient funds so that the compensation program can be administered by the superintendent in a fair and equitable manner. Although board members may provide input to the superintendent regarding their experiences with a given administrator, the superintendent alone should determine the evaluation and resulting compensation.

Few tasks of the superintendent create as much distance and isolation from subordinates as does the performance appraisal process. Poor performers usually resent the critical comments that they receive as part of the process and naturally believe that the superintendent is either supercritical or does not appreciate the difficulties that they encounter in trying to do their jobs. In some cases these negative feelings go beyond resentment to open anger and hostility. The task for the superintendent is to maintain a positive working relationship with the administrator whose performance is mediocre or unsatisfactory.

Even top performers are not as easy to supervise as one might imagine. In our experience the best administrators are demanding of themselves and tend to be perfectionists. These generally positive attributes cause these administrators to overemphasize any critical comments that are made during the evaluation conference and to discount the positive comments that are made. Such administrators also expect detailed explanations and supporting data regarding any areas for improvement that are proposed.

This leaves the majority of administrators, who are solid but not spectacular performers. Year after year these competent people are evaluated and consistently receive evaluations that are good, but not great. We have found it particularly challenging to maintain positive morale and motivation for this large proportion of administrators. Such people are working conscientiously and effectively, but never receive special recognition through the appraisal process.

The above paragraphs illustrate that performance appraisal, done correctly, is hard work. It is also a task that is not particularly appreciated by either the subordinates of the superintendent or by the school board. It is easy to see why many superintendents do not accord this critical task a higher priority. Nonetheless, a major theme of this book is that an effective performance appraisal program is an essential element in the operation of an excellent school district.

The entire administrative evaluation and compensation program should be reviewed periodically either as a matter of policy or contract. The relatively rapid turnover of administration personnel makes it imperative that ownership of the program by the administrators be assured through their periodic involvement in reviewing and modifying the existing program. The rapid turnover in school board membership offers another reason for a relatively frequent review of the program.

Two sample compensation programs based on the principles detailed above are provided on the following pages.

TECHNOMIC SCHOOL DISTRICT
Performance Appraisal Guidelines and Compensation Plan for the Technomic School District Administrators

A. Appraisal Purposes

The purpose of this appraisal program is to measure the effectiveness of the administrative personnel in the Technomic School District and to use such measurements in the establishment of a compensation plan for the district's administrators. The program represents the district's commitment to quality performance and an evaluation process that promotes both individual and program improvement. The results of the evaluation process will be utilized to identify and recognize quality supervision and administration, to promote and assist professional growth, and to provide for a continuing interchange of ideas and improvement of communication throughout the district.

B. Overview

The evaluation of each administrator will be a two-stage process. In the first stage the administrators will be evaluated on the performance of those tasks included in their job descriptions. This rating will constitute two-thirds of the annual summative evaluation. The remaining one-third will be determined by the administrator's performance in a predetermined set of Individual Administrative Objectives (IAOs) above and beyond the regular and expected tasks listed in the job description. The combination of these two ratings will be used annually, in accordance with the approved administrative compensation plan, to determine the salary increase for each Technomic School District administrator.

C. Job Description Performance Criteria

1. The job description of each Technomic School District administrator reflects normal expectations in the daily and yearly performance of duties.
2. Each administrator will meet yearly with his or her immediate superior to review and update the job description. The agreed upon descriptors will be the basis for an annual summative evaluation of job performance.
3. Each administrator will be evaluated by the superior to whom they immediately report according to the job description.
4. The performance level of each task in the job description will be evaluated according to the following scale:

	4 = Superior (Always)
Satisfactory	3 = Very Good (Almost always)
	2 = Good (Frequently)
	1 = Fair (Needs Improvement)
Unsatisfactory	0 = Unsatisfactory (Rarely or never)

5. The final rating for each administrator's Job Description Performance (JDP) will be determined by averaging the numerical ratings of the various categories to the nearest tenth. This rating will constitute two-thirds of the administrator's total annual summative evaluation.

D. Individual Administrative Objectives

1. Each administrator, working in cooperation with his or her immediate superior, will annually develop a minimum of three Individual Administrative Objectives (IAOs).
2. IAOs must be separate and distinct from the tasks listed in the administrator's job description.
3. Each administrator's IAOs will be reviewed annually by the superintendent to assure that a level of quality control exists for objectives across the organization. The immediate superior and the superintendent will determine the level of each employee's IAOs using the following criteria:

Scope— To what extent do the IAOs involve a major segment of Technomic School District's programs or personnel?

Complexity— To what extent do the IAOs require development of a detailed management plan with major tasks, completion dates, and people responsible for tasks?

Anticipated Impact— What will be the long-term and far-reaching impact on major segments of Technomic School District's personnel or programs?

Using these criteria, the immediate superior and superintendent will rate each proposed IAO according to the following scale:

4 = Superior
3 = Very Good
2 = Good
1 = Fair
0 = Unsatisfactory

Administrators will be given the opportunity to upgrade their IAOs after being advised of the superior's initial rating.
4. Each administrator's immediate superior shall annually determine the level of performance in each IAO and rate such performance as either "Satisfactory" or "Unsatisfactory." A "Satisfactory" rating in a Superior IAO earns 4 points; in a Very Good IAO, 3 points; and so on.
5. An annual average IAO rating for each Technomic School District administrator will be determined by dividing total points acquired by objectives attempted.
6. An administrator's average IAO rating shall constitute one-third of his or her overall annual summative evaluation.

E. Composite Annual Summative Evaluation

1. An administrator's annual summative evaluation will be determined by multiplying the average Job Description Performance rating by six and seven tenths (6.7), multiplying the average of the Individual Administrative Objectives rating by three and three tenths (3.3), adding the sums, then diving the sum by ten (10).

   For example, if Administrator Smith's average Job Description Performance rating was 3.2 and his Individual Administrative Objective rating was 3.5, his composite rating would be determined in the following manner:

$$\text{JDP rating} \quad 3.2 \times 6.7 = 21.44$$
$$\text{IAO rating} \quad 3.5 \times 3.3 = 11.55$$
$$\text{Sum} = 32.99 - 10 = 3.3$$

2. Composite ratings scale – Using the following scale, each administrator's composite score is converted into a summative rating:

Average Score	Summative Rating
3.6 or above	Superior
2.6 to 3.5	Very Good
1.6 to 2.5	Good
1.0 to 1.5	Fair
.9 or below	Unsatisfactory

F. Evaluation Schedule

1. By July 1, job descriptions are reviewed and updated.
2. By July 15, Individual Administrative Objectives (IAOs) proposed for the coming school year are submitted to the immediate superior and superintendent for review.
3. By July 22, the immediate superior has met with the administrators to review IAOs and return them for possible revision.
4. By August 1, IAOs for the coming school year are finalized.
5. In November and March, each administrator meets with his or her immediate superior to discuss progress on IAOs (written status reports should be sent to the superior prior to these meetings) and receives a formative evaluation report on job description task performance.
6. By May 1, each administrator submits a year-end status report of IAOs to his or her immediate superior and completes a self-evaluation after obtaining and reviewing a Summary Job Performance Assessment from his or her staff.
7. By May 15, each administrator meets with his or her superior to finalize the summative evaluation and rating of his or her JDP and IAOs. Information for this conference will include the administrator's self-evaluation, which incorporates staff assessment of performance, and the superior's observations.

G. Compensation Plan

The superintendent of schools and school board will annually review each administrative job description in determining base (minimum) salaries. The following factors will be considered:

1. Knowledge required—includes formal education, special training, and experience.
2. Supervisory responsibility—includes number of people supervised, complexity, and scope.
3. Accountability—includes immediate and long-range planning, freedom to act, execution of duties, and effects of job on end results.
4. Relationships with others—includes contact with the public, the students, the staff, and other administrators/supervisors.
5. Financial responsibility—includes development and control of budget and effects of error.
6. Leadership responsibility—includes initiative, stability, judgement, and motivating and directing others.
7. Communications responsibility—includes public and private information.
8. Job conditions—includes physical, mental and coordinative efforts, time spent on the job, problem solving, and decision making.
9. Organizational responsibilities—includes selection, training, and assignment of personnel.
10. Physical plant responsibility—includes layout, maintenance, and housekeeping.
   The minimum salary for each position is established by the board and adjusted as necessary due to changes in area cost-of-living conditions, job turnover or competitive salaries, changes in responsibility value, or other reasons as applicable.

TECHNOMIC SCHOOL DISTRICT
Administrator Performance Appraisal
Board Policy

The Technomic School District recognizes the importance of good supervision and evaluation of the educational programs and services it operates on behalf of the community. Because of this belief, the district supports the importance of a comprehensive evaluation and compensation program for its administrative staff to ensure that:

- A planned high-quality learning environment for the students of the district is maintained.
- District educational programs and services are being responsive to students.

- District administrators recognize their strengths and weaknesses.
- Annual goals and Individual Administrative Objectives (IAOs) are established to encourage and nurture individual strengths and minimize weaknesses for the improvement of administrative employees.
- Interaction occurs between the board and administrative personnel about the district management issues.
- Management of the district remains up-to-date and maintains high standards of performance.
- Administrators are compensated in a fair and equitable manner for work performed.
- Salaries in similar school districts are reviewed by the board of education prior to agreement renewal.

The superintendent of schools is responsible for the establishment of a program of administrative observation, evaluation, and rating.

TECHNOMIC SCHOOL DISTRICT
Administrator Performance Appraisal
Administrative Guidelines

The term *administrator* shall include the assistant superintendent for curriculum and instruction, director of personnel, director of business affairs, high school principal, assistant high school principal, middle school principal, assistant middle school principal, and the elementary school principals.

I. Administrator Evaluation

A. The district-approved Administrative/Supervisory Performance Appraisal Guidelines and Compensation Plan, accompanied by the state's standard rating form, shall be the official evaluation system for all state-certified administrators.
B. Administrator Performance Appraisal Guidelines
  1. *Job Performance* — Each administrator will be rated annually by his or her immediate superior using the board-approved job description for each administrative position.
  2. *Individual Administrative Objectives* — Each administrator will be required to undertake a minimum of three objectives each year. These objectives must be approved by the administrator's immediate superior and the superintendent.
    a. Objectives proposed for the coming school year must be submitted to the administrator's immediate superior by July 15 each year and approval must be finalized by August 1.

    b. Objectives can be reduced and/or waived on the basis of a recommendation by the administrator's superior and approval of the superintendent.

    c. Objectives not accomplished due to circumstances beyond the control of the administrator will be waived based on the recommendation of the immediate superior and approval of the superintendent.

3. *Performance Reviews* — Performance reviews of job performance and objectives will be held annually during November and March with a written progress report provided to each administrator.

4. *Final Evaluation* — A summative evaluation conference will be held annually by May 15. During this evaluation, a final rating will be given on job performance and on the Individual Administrative Objectives. The state's standard rating form will also be given at the May conference for state certified administrators. These ratings will determine salary adjustments as outlined in the agreement between the Technomic School District Board and administrators.

5. *Evaluation Appeal* — Administrators may disagree with any phase of their evaluation and may attach a written statement to their evaluation. Administrators may also appeal the evaluation to the superintendent. When the superintendent is the immediate superior, the employee may appeal directly to the board.

6. *Administrator Performance Appraisal Task Force* — A minimum of three board members, two administrators, and the superintendent will meet as required to review the Administrator Performance Appraisal Program and to arrive at a compensation plan. Meetings to discuss any aspect of the program can be requested by any party at any time.

II. Observation Requirement

A. Informal observations of the administrator's performance of his or her duties will be conducted on an ongoing basis. Following an informal observation, the observer may confer immediately with the employee or wait until the next scheduled performance review.

    Items observed informally that could affect the administrator's overall rating should be discussed with the administrator within ten days. A written record of the observation and the conference must be given to the administrator and placed by the observer in the administrator's file. The administrator may exercise his or her right to respond to any written report that may be given.

III. General Evaluation and Rating Procedures

A. The state standard rating form shall be the official Technomic School District rating form.

B. An overall rating of "satisfactory" or "unsatisfactory" shall be assigned as follows:
    Satisfactory = 1.0-4.0
    Unsatisfactory = .9 or below

C. As per state requirements, any rating less than .9 requires substantiation through documentation. This documentation will be those evaluations used in the Administrator Performance Appraisal Program (i.e., job performance and Individual Administrative Objectives).

D. Rating shall be done by or under the supervision of the superintendent. No unsatisfactory rating shall be valid unless approved by the superintendent.

E. An evaluation and rating conference must be held between the superior and the administrator within ten working days after the final observation associated with the rating, at which time the following is to occur:

  1. A thorough performance review is to be conducted, including a review of:
    a. Informal and formal observations
    b. Job performance
    c. Numerical assigned rating (see III.B)

  2. The state standard rating form, with supporting documentation attached, shall be signed by the superior, superintendent, and administrator, and a copy given to the administrator.

  3. The administrator may submit a written reply to the rating as soon as possible but not later than ten working days after the evaluation. This reply will be dated and attached to the rating.

F. Rating of less than .9 in either the job description or Individual Administrative Objectives shall be substantiated by anecdotal records and discussed with the administrator within five working days after the day of the final observation preceding the rating. The discussion may take place before or after the rating is approved by the superintendent. The five-day limitation may be extended only because of emergency or extenuating circumstances. (Items E.1 through E.3 apply.)

G. Copies of the completed rating forms will be furnished to the administrator with the original copy being filed in the superintendent's office. The original copy of the rating forms shall be considered the official copy.

H. The state standard rating form and documentation are due in the superintendent's office by May 15 each year.

IV. Unsatisfactory Rating

Prior to an administrator receiving an overall unsatisfactory rating, either A or B as defined below must have occurred:

A. Observations of the administrator's performance must have been conducted.

    1. Each observation is to be followed by a conference resulting in an observation/conference report.

    2. Specific suggestions to remediate weak areas in the administrator's performance must be detailed in the observation/conference report.

B. If there is information available to the superintendent that the administrator is in violation of the state school laws (causes for termination of contract), the following must have occurred:

    1. The administrator must have been informed, in writing, of the alleged violation by the superintendent.

    2. A conference between the administrator and superintendent must have occurred unless unforeseen circumstances prevented it.

    3. A written summary of the conference shall have been placed in the administrator's personnel file with a copy to the administrator and superintendent.

    4. For all other occurrences, B.1, B.2, and B.3 above apply.

    5. The outcome of the procedures listed above should be reflected on a state standard rating form.

V. Compensation Plans

The authority for developing administrative compensation plans results from laws passed by the state legislature.

A. *Meet and Discuss* — The law provides that if a majority of eligible administrators request in writing to meet with the employer regarding administrative compensation, the employer will meet and discuss in good faith prior to adopting an administrative compensation plan.

B. *Compensation Plan* — The plan must include, but does not have to be limited to

- a description of the program determining administrative salaries
- salary amounts and/or schedule
- a listing of fringe benefits

C. *Anti-Strike Law* — Administrative employees are subject to the "Public Employee Anti-Strike Law."

D. *Fact-Finding* — The Anti-Strike Law provides for an ad hoc three-member fact-finding panel to review and make findings and report those findings to the school employer for consideration.

    The State Department of Education has the statutory responsibility to appoint a third "neutral" member to the fact-finding panel in school disputes.

The program presented on the previous pages represents a system that fully employs all of the elements recommended in this book. The example to follow contains the same elements, although several of them are incorporated in a more basic form. Also, the particular manner in which the plan is written is a result of extensive involvement of administrators in the school system.

One additional element found in the following example involves a leadership skills document that is a separate factor in the performance appraisal process. This document was formulated as part of the program revision process of the administrators within the school district. While job descriptions identify which tasks are to be performed, the leadership skills component addresses the manner in which the administrator performs his or her tasks.

This document represents a revision of an existing evaluation and compensation program in the Octorara Area School District in Atglen, Pennsylvania.

Administrative Performance Appraisal Program for the Octorara Area School District

A. Purpose

The major purpose of the administrative performance appraisal is to improve the quality of administrative services provided to the students and staff of the school district. The appraisal program is designed to enhance communication between the superintendent and other administrators and to identify and concentrate upon those job responsibilities and skills that most directly contribute to a high-quality educational program.

B. Formal Program Elements

The annual performance appraisal of each administrator will be based upon the following three factors:

1. *Management by Objectives* — Each administrator will propose a small number of major objectives for each school year. These objectives should extend beyond the normal day-to-day requirements of the position and should consist of projects or programs to improve the educational program and/or the operation of the school. Specific objectives may be modified or deleted during the course of the year by mutual agreement between the administrator and the superintendent.

2. *Position Descriptions* — Each administrator will be evaluated in terms of his or her success in performing the tasks and obligations detailed in the position description. The position description enumerates the major functions and tasks of the administrator in terms of his or her position classification.

3. *Leadership Skills* — Each administrator will be evaluated in terms of his or her performance in relation to the leadership skills document developed by the administrative group. This list of skills represents

the findings of research and the consensus of opinion of our administrators regarding those skills and behaviors which are associated with successful school administration. It is not expected that all of these skills will be fully developed in any administrator at any given time.

C. Procedures

1. During the summer months, each administrator will meet with the superintendent to develop objectives for the next school year and to review items on the position description and/or the leadership skills list which are of interest or concern to either party. At this time, the superintendent will share any goals of the school board and/or the superintendent, which might influence the choice of objectives by the administrator.

2. A mid-year evaluation conference will be held to review progress on the administrator's objectives for the year and to discuss the position description and administrative skills list. The administrator shall prepare a written report detailing the status of his or her objectives and should be prepared to discuss selected items on the job description and/or leadership skills document that are of interest or concern to him or her.

3. In May of each year, a summary evaluation conference will be held between the superintendent and each administrator. A written report will be completed by the superintendent, which will address and consider administrator performance in terms of objectives for the year, position description, and leadership skills. The administrator should prepare for this meeting in the same manner as outlined above for the mid-year evaluation conference.

4. Each administrator will receive a summary evaluation of Superior, Very Good, Good, Satisfactory, or Unsatisfactory as a result of the total evaluation process. This evaluation will then be applied to the administrative compensation plan to determine salary for the next school year.

D. Supplementary Program Elements

1. It is considered to be good professional practice for an administrator to solicit opinions and feedback from his or her subordinates regarding the manner in which the administrator is fulfilling his or her role and job functions. It is expected that each administrator will periodically collect such information from subordinates in a manner and format to be chosen by each administrator. The administrator is neither required nor expected to share this information with his or her superior, but may do so if he or she so desires.

2. Each new administrator to the school district will be matched with an experienced Octorara administrator who will serve as mentor to the new administrator during his or her first year at Octorara. The superintendent will seek to identify experienced administrators as

mentors who are interested in serving in this capacity and who will be most likely to have a commonality of interests with the new administrator. In no case should the mentor be in a direct supervisory relationship to the new administrator.

3. Administrators new to the district should have scheduled conferences, in addition to regular evaluation conferences, to meet with the superintendent to discuss matters of interest to them regarding aspects of their position, management, and leadership of the Octorara Area School District and school administration in general. Such meetings with the superintendent should be scheduled periodically by the new administrator throughout his or her first year in the school district.

LEADERSHIP SKILLS
Octorara Administrator Appraisal System

A. Vision

Formulates his or her missions, influences events in the pursuit of a mission, and seeks creative approaches to the achievement of goals and objectives.

B. Organization

Planning: Assess needs, determines objectives, enlists the assistance of others, allocates resources, and develops realistic time schedules.
Delegation: Assesses the need for delegation; determines the person responsible; effectively communicates the tasks, authority, and timelines; and monitors progress as necessary.
Time Management: Plans, schedules, monitors, and adjusts his or her work in an effective and efficient manner and provides direction to subordinates.

C. Problem Solving

Problem Analysis: Evaluates situations to reach appropriate conclusions based upon the evidence at hand.
Judgment: Evaluates situations to reach logical conclusions based upon the evidence at hand.
Decision Making: Determines actions based upon available knowledge.

D. Task Completion

Carries out an objective, responsibility, or assignment in a timely fashion to a successful conclusion.

E. Evaluation

Systematically assesses outcomes of objectives, responsibilities, or assignments and uses feedback to implement plans for improvement.

F. Communication

Oral: Articulates issues in a clear, concise manner appropriate to the intended audience.
Written: Communicates issues in a clear, concise manner appropriate to the intended audience.
Listening: Demonstrates an understanding of and an interest in what is being stated.

G. Interpersonal Relationships

Motivation of Others: Encourages the initiative of others.
Facilitation: Moves toward mutual understanding of central issues and seeks consensus.
Sensitivity: Shows a sincere regard and respect for all individuals in the school community and promotes positive morale.
Flexibility: Examines issues objectively and is receptive to change when appropriate.
Stress Tolerance: Performs well under pressure and diffuses stress appropriately.

H. Loyalty

Demonstrates commitment to community, policies, and school and promotes the objectives of the school district and the needs of the students.

COMPENSATION PLAN FOR OCTORARA ADMINISTRATORS

1. The salary range for administrative positions in the Octorara Area School District will be determined by using a "control rate" that is based upon the current county average salary for a given administrative position.
2. The entry level for a given position will be set at 85 percent of the "control rate," the maximum for a given position will be set at 110 percent of the "control rate" for twelve-month positions.
3. A new "control rate" will be calculated for the upcoming school year by increasing the current "control rate" by the projected percentage increase in administrative salaries within the county for the next school year.
4. Using the above criteria, salary ranges for the 1991-92 school year have been calculated as indicated below based on a projected 7 percent increase in the 1990-91 "control rates."

Position	1990-91 Avg. Salaries	1991-92 Control Rate	1991-92 Salary Range
Director Curriculum/ Instruction (12)	60,958	64,615	54,923-71,077

Position	1990-91 Avg. Salaries	1991-92 Control Rate	1991-92 Salary Range
High School Principal (12)	60,958	64,615	54,923-71,077
Intermediate School Principal (12)	59,063	62,607	53,216-68,868
Elementary School Principal (12)	54,256	57,511	48,885-63,262
Director Special Services (12)	48,830	51,760	43,996-56,963
High School Assistant Principal (12)	48,571	51,485	43,762-56,634
Intermediate School Assistant Principal (12)	46,219	48,992	41,643-53,891

5. Salary ranges for other than twelve-month positions have been prorated to reflect ten or ten and a half months of work.
6. Salary ranges during the period of this agreement will be adjusted each year to reflect changes in the average salary for each position in the county. It is foreseeable that a ten/ten and a half month administrator's salary at maximum could fall below an Octorara teacher's salary at maximum at a later point in the life of this contract. If not corrected by an increased county average (the control rate), a committee of administrators and board members will meet to discuss adjusting salary ranges for affected positions.
7. Incumbent administrators will receive a base salary increase each year equal to the percentage increase in the "control rate" for the next school year.
8. Salaries for administrative personnel will be related to performance evaluation in the following manner:

Superior – base salary increase + 5 percent
Very good – base salary increase + 3.5 percent
Good – base salary increase + 2 percent
Satisfactory – base salary increase
Unsatisfactory – no increase

9. Movement to the top of the Octorara scale should take at least five years but not more than ten years.
10. Administrators at the top of their salary range will receive a base salary increase only. A special grant will be awarded, however, to personnel at the top of their scale who have been rated superior.
11. The control rate for the Director of Special Services will be equal to 90 percent of the control rate for an elementary principal.
12. The control rate for the Director of Curriculum and Instruction will be equal to the control rate for the high school principal.

Unsatisfactory Performance of a School Administrator

The primary goal of an administrative performance appraisal program is to impact positively on the performance of the vast majority of competent and conscientious school administrators. A secondary but vital purpose is to identify and ultimately dismiss those administrators whose performance is unsatisfactory and whose continued presence is harmful to students.

In its demands for accountability, the public is looking for positive results from school administrators above any other group. The public rightly views administrators as the group legally accountable for the quality of the schools and also is aware that they are, by public employment standards, well paid for the work that they do. Citizens are quick to remind us that they entrust us with two entities of great value to them, their children and their money.

As career public school administrators, we believe that proper management and leadership in the schools does make a difference. It therefore follows that poor leadership will have a negative impact on students. Thus, a superintendent must take seriously his or her obligation to either improve the performance of marginal administrators or take the proper steps to separate them from school district employment.

The most significant administrative positions from the standpoint of their direct effect on students are the superintendency and the building principalship. Securing effective leadership at the superintendency level is the most important responsibility of a school board. Guaranteeing good leadership at the building principal level is one of the major responsibilities of a superintendent. In their combined forty-five years of administrative experience, the authors have occasionally seen a good principal who, because of circumstances, has been unable to create a good school. However, we have never seen a case where a poor principal was able to either create or maintain a good school.

If our school district does in fact have one or more administrators who are indifferent and/or incompetent, how do we bring about their separation from school district employment in a fair and professional manner? The purpose of this section is to describe the procedures that should be followed in the dismissal process and to demonstrate the manner in which effective performance appraisal procedures can play a vital role in a dismissal situation.

Identifying the Unsatisfactory Administrator

The performance appraisal system described in this book is most useful in dealing with the more difficult cases of administrative incompetence. Obvious instances of administrative incompetence or malfeasance are relatively easy to deal with and do not require the detailed process that is the subject of this section. Examples of such easy dismissal calls would be the principal who runs away with his secretary or the assistant principal who steals money from the student activities fund. A third example would be a principal who has withdrawn so fully from his or her responsibilities that student discipline is totally lacking in the school and building operations are sliding toward chaos.

Each of the above scenarios, while all too common, are straightforward and should be quickly resolved by any competent superintendent and school board. In this discussion we will deal with the more difficult cases. For example, consider the high school principal who is so intimidated by his staff and the union that he will not properly supervise several teachers who are clearly incompetent. In addition, this principal has personally identified so closely with his school that he does not see and will not accept that academic achievement levels of his students are abysmally low. Finally, this person greets any attempt by the superintendent to address these issues as an attack, and he has conveyed this bunker mentality to his staff, thereby causing serious morale problems.

It is possible that all of the above can be true, and yet the principal could be perceived as competent by parents, students, and the school board. Teachers could be willing to acquiesce to his leadership since he does not pose a threat to their weaker members and does not make demands for positive change. How can a conscientious superintendent deal effectively with this situation?

The Progressive Discipline Process

Once a superintendent determines that the performance of an administrator may be seriously deficient, the concept of "progressive discipline" should be brought to bear on the problem. Progressive discipline can be defined as a process for ensuring that a systematic, fair, and thorough procedure is implemented for appraising the performance of a school employee. This same basic procedure should be utilized to deal effectively with poor performers at any level in the organization.

An effective progressive discipline plan should contain the following elements:

(1) *Identification of the problem or concern:* Is the supervisor reasonably certain that he or she is correctly perceiving the situation? Is the resolution of the problem within the responsibility and authority of the administrator? Is the perceived problem of concern to others such as teachers, students, parents, or other administrators?

(2) *Documentation of the problem:* Has the supervisor noticed a pattern of incidents relating to the area of concern? Has the supervisor begun to make anecdotal records relating to the problem? Has the problem been discussed either informally or during previous appraisal conferences with the administrator? Does the administrator in question perceive a problem to exist?

(3) *Development of a plan, with timelines, to address the problem:* Has the supervisor explicitly expressed his or her concerns about the problem to the administrator? Have the supervisor and the employee jointly agreed upon a plan of action for correcting the situation? Have benchmarks been agreed upon for reevaluating the situation in question?

(4) *Provision of assistance to correct the problem:* Does the administrator possess the tools needed to deal with the problem? Have staff development opportunities been made available to the administrator? Has the superintendent fulfilled his or her responsibilities as a coach and mentor to assist the administrator? Does the administrator have the staff and material resources necessary to address the problem?

(5) *Periodic assessments of progress toward resolution of the problem:* Have periodic meetings been held, monthly or quarterly, between the superintendent and the administrator to discuss progress toward achievement of goals? Have the results of these meetings been documented? Has the administrator been required to provide written evidence supporting the steps he or she has taken with respect to the problem? Has the supervisor provided written responses to the reports of the administrator?

(6) *Official unsatisfactory rating in accordance with local policy and state law:* Does the written documentation relating to an unsatisfactory rating address legal requirements in every detail? Does the written commentary accompanying the rating explicitly state that a second unsatisfactory rating will lead to a recommendation for dismissal? Does the written evaluation explicitly list those aspects of performance that are unsatisfactory, and does the statement also contain specific suggestions for correcting the problem? Are the additional resources and other assistance provided to the administrator listed in the written document? Has the administrator signed the unsatisfactory rating indicating awareness of the rating and the reasons for it? Has the administrator been given an opportunity to contest the rating in writing?

(7) *Sufficient time for remedial action:* Has a period of at least a few months been provided for the administrator to undertake corrective action? Have continuous periodic meetings, with documentation, been held between the administrator and the superintendent to review progress relating to the problem? Do these documents clearly state any continuing concerns that the superintendent might have?

(8) *Second unsatisfactory rating and recommendation for dismissal:* Is the supervisor able to document that the administrator has failed to adequately address the problems enumerated in the first unsatisfactory rating? Can the supervisor demonstrate that the level of assistance outlined in the first rating has in fact been provided? Has the rating been signed by the administrator and has he or she been given a chance to contest the rating in writing?

In most states the school board will ultimately decide whether or not the superintendent's unsatisfactory rating of the administrator will be upheld and the administrator will be dismissed. Even a dismissal by the school board, however, does not preclude the possibility that the case will be appealed to the state department of education or through the court system. Whatever the situation, a superintendent who faithfully follows the progressive discipline procedures outlined above can approach any forum that reviews the case with confidence.

A major benefit of a thorough progressive discipline program is that the administrator involved in the process will almost always voluntarily resign rather than contest a dismissal recommendation that is solidly documented and is the result of a process that was fair and equitable to the administrator. In many cases, the administrator will respond rather early to the evidence of deficiencies and his or her unwillingness or inability to correct the problem. In this situation the superintendent is in the role of counseling the administrator out of public school administration. This development allows the administrator to resign voluntarily and with dignity. The hallmark of an effective progressive discipline system is that it seldom needs to be fully implemented.

Performance Appraisal Guidelines and Progressive Discipline

Now that we have fully discussed the procedures to be followed in the implementation of a progressive discipline system, we would like to review the ways in which the performance appraisal guidelines provided in this book can complement that system. The most basic task to be addressed in the area of performance appraisal is to identify the actual job that is to be assessed. A thorough and accurate job description for each position in the organization is therefore a necessary first step in a progressive discipline program.

The performance appraisal program recommended in this book utilizes the job description as the fundamental document in assessing job performance. These job descriptions should be formally approved by the school board and should be viewed as a regular part of the appraisal process by all administrators. Thus when the superintendent makes reference to an area of concern relating to the job description, the administrator in question should not be surprised to learn that the particular item in question is one of his or her job responsibilities. A long-term employee, in fact, will almost certainly have discussed this particular job responsibility with the superintendent at some previous evaluation conference.

The Individual Administrative Objectives can also be used effectively as part of a progressive discipline process. The supervisor can require that the administrator develop one or more objectives for the year relating directly to the areas of concern regarding the performance of the administrator. These objectives will be written by the administrator in terms of the activities to be attempted and the documentation that will be used to evaluate the achievement of the objective.

By requiring the administrator to develop his or her own strategies for resolving a problem, the superintendent is maximizing the chance that the administrator will take ownership for the problem. This approach also is fairest to the administrator since he or she is able to choose the methods whereby the problem can be resolved. Listing the problem as one of the administrator's four or five major objectives for the year emphasizes the importance that the supervisor attaches to the resolution of the problem.

The periodic meetings required by the performance appraisal program outlined in this book also provide the general structure around which to schedule additional meetings to meet the requirements of a progressive discipline process. An administrator who has worked with this type of performance appraisal program for several years will be no stranger to the process of meeting periodically with a supervisor to discuss performance levels. The superintendent will also have benefited from the many opportunities that he or she will have had over the years to critically discuss performance with many administrators on many occasions. While conducting critical performance reviews is never pleasant, it does become somewhat easier with practice.

In concluding our discussion of the topic of dismissal of administrators, we would like to reiterate our conviction that an effective performance appraisal program will greatly diminish the need to initiate dismissal proceedings against administrative personnel. The constant communication and critical review of performance implied by the system will minimize the number of situations that will require consideration of dismissal. Also, an effective appraisal system will help administrators to better recognize their own deficiencies early on so that they can correct them or voluntarily move on to other career options.

CHAPTER NINE

From Theory into Practice

CONVERTING the theoretical discussion of performance appraisal contained in the previous chapters into a practical reality is the subject of this chapter. Performance appraisal is a topic that will have a direct and personal interest for all members of any management team. Thus, to an extent not found in discussions of other topics, the decision to review performance appraisal for administrators will bring forth expressions of deep-seated feelings and emotions as well as sound judgments based on rational thinking and past experience.

Some members of the management team will be open to a proposed review of performance appraisal and will look upon such a discussion as an opportunity for further professional growth and development. Less secure administrators, on the other hand, may view the process as a threat to their security, status, and compensation level. The challenge to the superintendent during this process is to capitalize on the contributions of those administrators who are open to the process while at the same time finding ways to allay the fears and gain the support of the more reluctant participants.

A Timeline for the Review Process

A project on the magnitude of revising an administrative performance appraisal program will require the greater part of a school year to address adequately. Ample time must be allowed for administrators to grapple with the idea of such a potentially significant change in their work environment. In addition, time must be allowed to research and discuss various models of administrative performance appraisal.

155

Once a general outline of a new appraisal process has been formulated, frequent opportunities must be provided for reflection and discussion among all levels of the administrative team. Every attempt should be made to address the suggestions and concerns voiced by affected personnel. A final report should be prepared for school board approval only after ample time has been provided to secure consensus from the administrative team regarding the principal elements of the proposed plan.

Reviewing the Existing Program

The first step in reviewing the performance appraisal program in a school district should be a critical analysis of the existing program. If some period of time has passed since the inception of the current program, it is likely that the program is not being fully implemented as designed. There is a tendency in any organization for written protocols and procedures to be subtly modified over time by those who are actually implementing a policy or program.

Also, the superintendent and other members of the administrative team may have changed over time so that few of the people who originally designed the program are currently employed by the school district. In reality, newer members of the administrative staff may have only a vague understanding of the rationale for the program or the manner in which it was designed to operate. Thus this initial review of the existing program will serve as a staff development exercise on the topic of administrative evaluation for the entire administrative team.

Unless the size of the administrative team is prohibitive, we would recommend that all members of the administration be included in the process of reviewing the rationale for and the strengths and weaknesses of the current system. The data gathered through this process of internal analysis are a necessary but not sufficient basis for approaching the task of developing a new appraisal program.

Searching for the State of the Art

The next stage in the process is to examine current thinking and practice in the area of performance appraisal in the field of education. Sources of such information include state and national administrative and school board organizations, books, and journal articles on the topic, as well as recognized experts on the subject from regional universities or school districts.

An outside consultant or facilitator might be helpful at this point to work with the administrative group as it evaluates other programs and approaches. Such an objective participant should facilitate a freer and more open exchange among administrators as they begin to grapple with the components of a new administrative evaluation process for their school district. It is at this stage that administrators will be asked to consider elements for their program that will be unfamiliar to them. An outside consultant is in a good position to discuss the pros and cons of these new approaches in a nonthreatening and collegial atmosphere.

The role of the superintendent should be to offer support and encouragement to the discussion process without attempting to force his or her viewpoints on the group. It is better to accept fewer changes in the program that are acceptable to the group rather than to impose a process that will be threatening to some members of the administrative team.

One element that administrators might find unfamiliar is the concept of self-evaluation as a formal component of the performance appraisal process. Although administrators will be familiar with the practice of being asked for their opinions of their performance as part of the evaluation conference, they will be less familiar with the suggestion that they should arrive at the conference with a formal written appraisal of their own performance in the areas of job description and individual administrative objectives.

An experienced supervisor will reflect that in his or her experience some subordinates prepare such a written self-appraisal as a matter of course, while others arrive at the evaluation conference prepared to listen and respond to direct questions but with little inclination to directly reflect on their own performance. This latter group of administrators consists of individuals who need to be convinced of the inherent advantages of critical self-appraisal in the presence of a supervisor. They need to be able to perceive that the supervisor will play the role of coach in helping them to analyze past performance in an effort to enhance future effectiveness.

The experience of one of the authors in incorporating a self-evaluation component into a revised appraisal system is instructive.

Some administrators in the district were accustomed to presenting detailed reports about progress on their IAOs while others were not. It was not at all difficult to convince all administrators that detailed written reports on achievement of objectives was both reasonable and helpful to the administrators.

Asking administrators to arrive at an evaluation conference prepared to critically evaluate their performance on their job descriptions and to formally comment on their level of attainment on the various leadership skills dimensions of the appraisal system was more problematic. Several administrators were simply not comfortable with the idea of providing potentially self-critical written commentary on their own performance.

After considerable discussion by the group, it was decided that each administrator would come to the evaluation conference prepared to verbally discuss several items from both his or her job description and the leadership skills document. Likewise, the superintendent would select several items from these two documents for discussion during the conference.

While this approach might be faulted by some as not being sufficiently formalized, everyone involved did accept the proposition that both the supervisor and the supervisee should critically review the performance of the administrator in terms of his or her job description and leadership skills. This seemed to the author involved to be a significant step forward for his district at that particular time.

A second concept likely to require significant discussion is the notion that subordinates should have a formal role to play in the appraisal process. The less secure the administrator, the more adamant will be his or her objection to evaluation by subordinates. It is therefore critical that the superintendent affirm repeatedly that these subordinate evaluations will not be shared with the supervisor.

The administrators must also realize, however, that they are free to share some information from these subordinate evaluations with their supervisor should they desire to do so. The point should also be made that permitting the administrator to voluntarily share this information with the supervisor provides an opportunity for the administrator to contribute some positive data to the appraisal conference that might not otherwise be available to the supervisor.

Even the negative feedback that an administrator may receive from subordinates can be put to good use. The administrator can consider those areas of his or her performance that are not rated highly and develop plans of action designed to improve performance in the future. Thus this subordinate evaluation process can be used solely by the administrator as a way to improve performance over time in ways that will ultimately reflect positively on his or her evaluation by the supervisor.

Incorporating an evaluation by subordinates into the performance appraisal process is perhaps the most difficult challenge in developing a new evaluation program. Only the most secure administrators will greet this concept with enthusiasm. During his superintendency one of the authors was able to help formulate an approach to this concept that satisfied the concerns of the administrators and yet firmly established the principle of subordinate involvement in the evaluation process within the school district.

Once the matter was thoroughly discussed there was agreement, at the theoretical level, that such subordinate input would be useful. In discussing the frequency of such an involvement process, there was concern that a fixed schedule might require subordinate input at an inappropriate time. The times immediately after a teacher strike or immediately following a particularly stormy encounter between a principal and his or her staff were given as examples. Thus it was ultimately decided that the appraisal program would require periodic subordinate input, with the exact timing of such an exercise left to the discretion of the individual administrator.

A second issue related to subordinate input concerned the forms and procedures that would be used. Here again it was ultimately decided that the forms and procedures would be left to the discretion of the administrator. Although this decision will undoubtedly dilute the effectiveness of the procedure in some instances, the acceptance of the concept and its inclusion in the formal appraisal program were both important accomplishments. It is also noteworthy that by structuring the subordinate input element in these ways the concept was unanimously endorsed by all members of the administrative team.

Introducing New Performance Appraisal Concepts

Both of the above concepts will be considered sufficiently radical or threatening to some administrators that the superintendent should allow a period of several months with several opportunities to discuss the ideas before formally calling for their consideration as a part of a revised performance appraisal program. As reluctant members of the administrative team begin to see that the purpose of a revised appraisal program is the professional development of them as individuals, they will become more open to incorporating elements into the program designed to promote such growth and development.

Once the concepts new to the group have been developed to a certain point, a good strategy might be to divide the group into subcommittees to further refine the major components of the evolving appraisal program. These subcommittees should not include the superintendent but should contain membership from the various organizational levels and position levels, including central office personnel.

These subcommittee meetings will allow administrative personnel to interact in an informal manner with their administrative colleagues and will allow them to discuss elements of the proposed system in a non-threatening context. As this process moves forward, the groups will tend to agree on fairly high-risk evaluation elements that might not have survived in the context of more formal and more constrained large group meetings with the superintendent or school board present.

In one school district where this subcommittee process was used, the administrators expressed great satisfaction in having this opportunity to meet informally with colleagues to discuss the essentials of their work and how their performance of these essential tasks should be evaluated. These administrators found this collaborative exercise to be a powerful method for team building and for gaining a better understanding of one another as fellow administrators. These positive outcomes were particularly pronounced among the newer administrators in the district.

Once the major elements of the new system have been refined by the subcommittees, it is time to discuss each of the elements again with the total group. Each subcommittee should present and defend its own report and modifications should be made on the basis of group input. The superintendent should be a participant in these discussions but should in no way dominate the discussion or attempt to force decisions with which the group is uncomfortable. As facilitator of the group, the superintendent should seek to develop consensus on each of the major issues under discussion.

School Board Involvement in the Process

We believe strongly that the appraisal program should be fully developed and supported by the administrative personnel. There are two reasons for this conviction. The first is that the administrative group has the greatest stake in the program since their evaluations and compensation are dependent upon it. The second reason is that the administrators have a longer term interest in the program than either the superintendent or the school board. The average tenure of superintendents in a school district is less than five years, while school board members tend to have similarly short-term involvements with the school district. Administrators, on the other hand, typically serve in the same district for ten to fifteen years or more.

Once a revised program has been endorsed by both the administrative group and the superintendent, it is time to involve the school board directly in the process. The rationale for the program as well as the instruments and operating procedures should be presented to the total school board at an evening meeting or retreat setting dedicated solely to a discussion of the administrative team, its evaluation, and compensation.

The school board should have the opportunity to make further modifications to the proposed system as a result of this meeting. If the superintendent has done his or her work well, the program as initially presented should generally reflect the thinking of the school board and only relatively minor changes should be necessary at this point. After some further refinement and consideration by the school board over a period of one or two months, the new program should be ready for adoption by the school board and implementation by the superintendent.

Revision of School Board Appraisal Program

In Chapter Seven we reviewed in detail the steps to be taken in revising the existing performance appraisal process for the school board itself. The logical time to initiate this process with the school board would be immediately following the school board's consideration and approval of a new performance appraisal system for its administrators.

As a result of reviewing the principal features of the new administrative evaluation program, board members will be knowledgeable regarding current good practice in the area of performance appraisal. They will have a ready appreciation for the underlying rationale for the performance appraisal process. This level of experience with performance appraisal systems should make board members receptive to the idea that they should involve themselves in an appraisal process similar to the one recently ratified for their immediate subordinates.

The superintendent must play a leadership role in facilitating the development of a revised performance appraisal system for the school board. The best vehicle for this would be for the school board to establish a committee to develop such a system with the superintendent as a consultant to the committee.

This board committee could then follow the same procedures for reviewing the existing program and exploring alternative programs that have been described above for the development of the administrative performance program. This process should be simplified by the fact that board members will have available for study the new program that they recently approved for administrative personnel.

A special work session or retreat setting would be an appropriate context in which to present a proposed new program to the entire board. As was true for the administrators, all board members should feel comfortable with the components of the program before a decision is made to implement the revised system. The approved system should be formally approved at a board meeting so that adhering to the timelines and other requirements of the plan will be a matter of policy.

Publicizing the New Performance Appraisal Program

Throughout this book we have focused on the direct benefits of a comprehensive performance appraisal program for the administrators and for the school board. We should also consider the positive impact that such programs can achieve both with the professional staff and with the external publics.

A building administrator, for example, can use the elements of his or her appraisal process as a means of periodically reviewing with faculty his or her major objectives for the year as well as items on the job description to which he or she is giving special emphasis. These periodic reminders to staff members that their leader is goal-oriented and is critically evaluating his or her own performance will promote an atmosphere of collaboration and collective purpose. Also, we should not underestimate the positive model that an administrator willing to engage in critical self-analysis can have in promoting similar behavior among his or her subordinates.

The implementation of a rigorous performance appraisal plan for administrators as well as the school board should be widely and frequently publicized to the general public. Citizens often view school administrators as government bureaucrats with all of the negative connotations that the term can imply. Chief among these negative stereotypes is the conviction that government workers are not held accountable for their performance and that there are no incentives for them to perform well.

All of us as superintendents or school board members have faced critics at board meetings who questioned the number of administrators, their competence, and their salaries. In all but the most affluent communities, school administrators are paid far better than the average taxpayer. They are also performing a function that is a mystery to the typical taxpayer and on which he or she places little value.

We must be able to face these critics armed with job descriptions that fully explain the complexities of the administrative profession. We must be able to demonstrate that we have initiated a rigorous and effective system for evaluating the quality of administrative performance. We must also be able to demonstrate that administrative salary increases are awarded on merit and that such raises are withheld in cases of poor performance. In short, an integrated administrative appraisal and compensation program will help us to achieve both the image and reality of effective school district leadership.

The performance appraisal concepts that we have explored in this book will lead to a high level of accountability for administrators, will provide a systematic process for continually evaluating the level of administrative performance, and will directly relate compensation to the quality of performance. A district that adopts the program recommended in this book is institutionalizing a mechanism to ensure dynamic and effective leadership for a school district. Every effort should be made to publicize this fact so that the community will properly adopt a more positive attitude toward its administrative and school board leadership.

Implementing the New Performance Appraisal System

The actual implementation of the revised performance appraisal program should be a primary focus of the superintendent during its first year of use. If the revision process has been successful, the new program will contain several elements with which administrators will be unfamiliar and which may be the cause of some anxiety. Administrators may also need some help with correctly initiating some procedures in the process that are new to them.

The superintendent should begin the first year of implementation by carefully reviewing with administrators each element of the program. Particular attention should be given to elements that will require the involvement of other people, such as the process for securing subordinate input in administrative evaluation. The superintendent will need to closely monitor such aspects of the program to make certain that an individual administrator does not misinterpret or misapply a new element in the process.

The long-term success of the revised program also dictates that the superintendent make some allowances during the first year regarding the quality of the early attempts by administrators to implement the system. As a good coach, the superintendent will want to support the administrators through some guided practice of their new behaviors. In this manner the new system will be a positive experience for administrative personnel and the positive potential of the new system will be realized more fully with each passing year.

The implementation of this total process should result in substantial improvement in existing practices for appraising all levels of the administration, including the superintendent, as well as appraisal of the school board itself. The involvement of all levels of school system leadership in this endeavor will also reinforce the concept of the management team and will enhance collaborative relationships among members of the leadership cadre.

165